The
Sports
Gene

CURRENT

The

Sports

Gene

Inside the Science of Extraordinary
Athletic Performance

David Epstein

CURRENT

CURRENT
Published by the Penguin Group
Penguin Group (USA) Inc., 375 Hudson Street,
New York, New York 10014, USA

USA | Canada | UK | Ireland | Australia | New Zealand | India | South Africa | China

Penguin Books Ltd, Registered Offices: 80 Strand, London WC2R 0RL, England
For more information about the Penguin Group visit penguin.com

LIBRARY OF CONGRESS CATALOGING IN PUBLICATION DATA
Epstein, David J.
   The sports gene : inside the science of extraordinary athletic
performance / David Epstein.
      pages cm.
   Includes bibliographical references and index.
   ISBN 978-1-59184-511-9
  1. Sports—Physiological aspects.   2. Human genetics.   I. Title.

   RC1235.E58 2013
   613.7'1—dc23   2013013443

Printed in the United States of America
10   9   8   7

Designed by Alissa Amell · Set in ITC New Baskerville

For Elizabeth, my very own MC1R gene mutant

CONTENTS

Contents

The
Sports
Gene

In Search of Sports Genes

Micheno Lawrence was a sprinter on my high school track team. The son of Jamaican parents, he was short and doughy and his bulging paunch poked at the holes of his marina, the mesh top that some Jamaicans on the team wore to practice. He worked at McDonald's after school, and teammates joked that he partook too often in the product. But it didn't stop him from being head-whipping fast.

A mini-diaspora in the 1970s and '80s brought a stream of Jamaican families to Evanston, Illinois, which helped make track and field a popular sport at Evanston Township High School. (Consequently, our team won twenty-four consecutive conference titles from 1976 to 1999.) As outstanding athletes are wont to do, Micheno referred to himself in the third person. "Micheno got no heart," he would say before a big race, meaning that he had no sympathy when vanquishing his competitors. In 1998, my senior year, he blasted from fourth place to first on the anchor leg of the 4×400-meter relay to win the Illinois state championship.

We all knew an athlete like that in high school. The one who made it look so easy. He was the starting quarterback and shortstop, or she was the all-state point guard and high jumper. *Naturals.*

Or were they? Did Eli and Peyton Manning inherit Archie's quarterback genes, or did they grow up to be Super Bowl MVPs because they were raised with a football in hand? Joe "Jellybean" Bryant clearly passed his stature to his son, Kobe, but where does that explosive first step come from? What about Paolo Maldini, who captained AC Milan to a Champions League title forty years after his father, Cesare, did the same? Did Ken Griffey Sr. gift his boy with baseball batter DNA? Or was the real gift that he raised Junior in a baseball clubhouse? Or both? In 2010, in a sporting first, the mother/daughter pair of Irina and Olga Lenskiy made up half of Israel's national team in the 4×100-meter relay. The speed gene *must* run in that family. But is there even such a thing? Do "sports genes" exist at all?

In April 2003, an international consortium of scientists announced the completion of the Human Genome Project. Following thirteen years of toil (and 200,000 years of anatomically modern man), the project had mapped the human genome; all 23,000 or so regions of DNA that contain genes had been identified. Suddenly, researchers knew where to begin looking for the deepest roots of human traits, from hair color to hereditary disease and hand-eye coordination; but they underestimated how difficult the genetic instructions would be to read.

Imagine the genome as a 23,000-page recipe book that resides at the center of every human cell and provides directions for the creation of the body. If you could read those 23,000 pages, then you would be able to understand everything about how the body is made. That was the wishful thinking of scientists, anyway. Instead, not only do some of the 23,000 pages have instructions for many different functions in the body, but if one page is moved, altered, or torn out, then some of the other 22,999 pages may suddenly contain new instructions.

In the years following the sequencing of the human genome, sports scientists picked single genes that they guessed would influence athleticism and compared different versions of those genes in small groups of athletes and nonathletes. Unfortunately for such studies, single genes usually have effects so tiny as to be undetectable in small studies. Even most of the genes for easily measured traits, such as height, largely eluded detection. Not because they don't exist, but because they were cloaked by the complexity of genetics.

Sluggishly but surely, scientists have begun to abandon the small, single-gene studies and steer the scientific ship toward new and innovative methods of analyzing how genetic instructions function. Couple that with the efforts of biologists, physiologists, and exercise scientists to discern how the interplay of biological endowments and rigorous training affects athleticism, and we're starting to tug at the threads of the great nature-versus-nurture debate as it bears on sports. That necessarily involves trekking deep into the bramble patches of sensitive topics like gender and race. Since science has gone there, this book will too.

The broad truth is that nature and nurture are so interlaced in any realm of athletic performance that the answer is always: it's both. But that is not a satisfactory endpoint in science. Scientists must ask, "How, specifically, might nature and nurture be at work here?" and "How much does each contribute?" In pursuit of answers to these questions, sports scientists have trundled into the era of modern genetic research. This book is my attempt to trace where they have gone and to examine much of what is known or haggled over about the innate gifts of elite athletes.

In high school, I wondered whether Micheno and the other children of Jamaican parents who made our team so successful might carry some special speed gene they imported from their tiny island. In college, I had the chance to run against Kenyans, and I wondered whether endurance genes might have traveled with them from East Africa. At the same time, I began to notice that a training group on my team could consist of five men who run next to one another, stride for stride, day after day, and nonetheless turn out five entirely different runners. How could this be?

After my college running career ended, I became a science graduate student and later a writer at *Sports Illustrated*. In researching and writing *The Sports Gene* I had the chance to blend in the petri dish of elite sports what initially seemed to me to be wholly separate interests in athleticism and science.

The reporting of this book took me below the equator and above the Arctic Circle, into contact with world and Olympic champions, and with animals and humans who possess rare gene mutations or outlandish physical traits that dramatically influence their athleticism. Along the way, I learned that some characteristics that I assumed were entirely voluntary, like an athlete's will to train, might in fact have important genetic components, and that others that I figured were largely innate, like the bullet-fast reactions of a baseball batter or cricket batsman, might not be.

Let's start there.

1

Beat by an Underhand Girl

The Gene-Free Model of Expertise

The American League team was deep in a hole, and National League slugger Mike Piazza was up to bat. So they called for the ringer.

Sauntering past a phalanx of the world's best hitters, Jennie Finch strode toward the sun-drenched infield, her flaxen hair blazing in the clear desert light. For the previous twenty-four years, the Pepsi All-Star Softball Game had been an event contested by Major League Baseball players only. The crowd thrummed with excitement as the 6'1" Team USA softball ace reached the pitcher's mound and curled her fingers around the ball.

It was a temperate day in Cathedral City, California; 70 degrees in the replica of one of America's own sports cathedrals. The three-quarter-scale imitation of the Chicago Cubs' Wrigley Field was faithful in its ivy-covered outfield walls. Even Wrigleyville's brick apartment buildings were there, in the desert at the foot of the Santa Rosa Mountains, depicted on near-life-size vinyl prints created from photographs of Chicago.

Finch, who in a few months would win a gold medal at the 2004 Olympics, had originally been invited only as a member of the American League coaching staff. That is, until the American League stars went down 9–1 in the fifth inning.

No sooner did Finch arrive at the mound than the defensive players behind her sat down. Yankees infielder Aaron Boone took his glove off, lay down in the dirt, and used second base for a pillow. Texas Rangers All-Star Hank Blalock took the opportunity to get a drink of water. They had, after all, seen Finch pitch during batting practice.

As part of the pregame festivities, a raft of major league stars had tested their skill against Finch's underhand rockets. Thrown from a mound forty-three feet away, and traveling at speeds in the upper-60-mph range, Finch's pitches take about the same time to reach home plate as a 95-mph fastball does from the standard baseball mound, sixty feet and six inches away. A 95-mph pitch is fast, certainly, but routine for pro baseball players. Plus, the softball is larger, which should make for easier contact.

Nonetheless, with each windmill arc of her arm, Finch blew pitches by the bemused men. When Albert Pujols, the greatest hitter of a generation, stepped forward to face Finch during pregame practice, the other major leaguers crowded around to gawk. Finch adjusted her ponytail nervously. A wide smile stole across her face. She was exhilarated, but also anxious that Pujols might hit a line drive right back at her. A silver chain dangled over his expansive chest, his forearms as wide as the barrel of the bat. "All right," Pujols said softly, indicating he was ready. Finch rocked back, and then forward, whipping her arm in a giant circle. She fired the first pitch just high. Pujols lurched backward, startled at what he saw. Finch giggled.

She unleashed another fastball, this time high and inside. Pujols spun defensively, turning his head away. Behind him, his professional peers guffawed. Pujols stepped out of the batter's box, composed himself, and stepped back in. He twisted his feet into the dirt, and stared back at Finch. The next pitch came right down the middle. Pujols uncoiled a violent swing. The ball sailed past his bat, and the spectators hooted. The next pitch was way outside, and Pujols let it go. The one after that was another strike, and Pujols whiffed again. With one strike remaining, Pujols moved all the way to the back of the batter's box and dug in, crouching low in his stance.

Finch rocked, and fired. Pujols missed, badly. He turned and walked away, toward his tittering teammates. Then he stopped, bewildered. Pujols turned back to Finch, doffed his cap, and continued on his way. "I don't want to experience that again," he later resolved.

So the defensive players behind Finch had good reason to sit down in the field when she entered the live game: they knew there would be no hits. Just as she had during the pregame practice, Finch struck out both hitters she faced. Piazza struck out on three straight pitches. San Diego Padres outfielder Brian Giles missed so badly on the third strike that his momentum spun him through a pirouette. And then Finch returned to her role as a ceremonial coach. She was, though, not nearly finished befuddling major leaguers.

In 2004 and 2005, Finch hosted a regular segment on Fox's *This Week in Baseball* in which she would travel to major league training camps and transform the best baseball hitters in the world into clumsy hacks.

"Girls hit this stuff?" asked an incredulous Mike Cameron, the Seattle Mariners outfielder, after he missed a pitch by half a foot.

When seven-time MVP Barry Bonds saw Finch at the Major League All-Star Game, he walked through a throng of media so that he could talk trash to her.

"So, Barry, when do I get to face the best?" Finch asked.

"Whenever you want to," Bonds replied, confidently. "You faced all them little chumps. . . . You gotta face the best. You can't be pretty and good, and not face another handsome guy who's good," Bonds said, simultaneously flirting and unfurling his peacock feathers. Bonds then told Finch to bring a protective net when she was ready to face him, because "you're going to need it with me . . . I'll hit you."

"There's only been one guy who touched it," Finch replied.

"Touch it?" Bonds said, laughing. "If it comes across that plate, believe me, I'ma touch it. I'ma touch it *hard*."

"I'll have my people call your people and we'll set it up," Finch told him.

"Oh, it's on! You can call me direct, girl," Bonds said. "I take my challenges *direct* . . . we'll televise it too, on national television. I want the world to see, everybody to see."

So Finch traveled to face Bonds—this time without fans and other media around—and the tune of his raillery quickly changed. Bonds watched several pitches fly by, and insisted that the cameras not film him. Finch shot pitch after pitch past Bonds, as his on-looking teammates pronounced them strikes. "That's a ball!" Bonds pleaded, to which one of his teammates replied, "Barry, you've got twelve umpires back here." Bonds watched dozens of strikes go by without so much as a swing. Not until Finch began to tell Bonds what pitches were coming did he tap a meek foul ball that rolled to rest a few feet away. Bonds implored Finch, "Go on, throw the cheese!" She did, and blew it right past him.

When Finch subsequently visited Alex Rodriguez, the reigning MVP, Rodriguez watched over Finch's shoulder as she threw warm-up pitches to one of his team's catchers. The catcher missed three of the first five throws. Seeing that, Rodriguez, to Finch's disappointment, simply refused to step into the batter's box. He leaned forward and told her: "No one's going to make a fool out of me."

For four decades, scientists have been constructing a picture of how elite athletes intercept speeding objects.

The intuitive explanation is that the Albert Pujolses and Roger Federers of the world simply have the genetic gift of quicker reflexes that provide them with more time to react to the ball. Except, that isn't true.

When people are tested for their "simple reaction time"—how fast they can hit a button in response to a light—most of us, whether we are teachers, lawyers, or pro athletes, take around 200 milliseconds, or one fifth of a second. A fifth of a second is about the minimum time that it takes for the retina at the back of the human eye to receive information and for that information to be conveyed across synapses—the gaps between neurons that take a few milliseconds each to cross—to the primary visual cortex in the back of the brain, and for the brain to send a message to the spinal cord that puts the muscles in motion. All this happens in the blink of an eye. (It takes 150 milliseconds just to execute a blink when a light is shined in your face.) But as quick as 200 milliseconds is, in the realm of 100-mph baseballs and 130-mph tennis serves, it is far too slow.

A typical major league fastball travels around ten feet in just the 75 milliseconds that it takes for sensory cells in the retina simply to confirm that a baseball is in view and for information about the flight path and velocity of the ball to be relayed to the brain. The entire flight of the baseball from the pitcher's hand to the plate takes just 400 milliseconds. And because it takes half that time merely to initiate muscular action, a major league batter has to know where he is swinging shortly after the ball has left the pitcher's hand, well before it's even halfway to the plate. The window for actually making contact with the ball, when it is in reach of the bat, is 5 milliseconds, and because the angular position of the ball relative to the hitter's eye changes so rapidly as it gets closer to the plate, the advice to "keep your eye on the ball" is literally impossible. Humans don't have a visual system fast enough to track the ball all the way in. A batter could just as well close his eyes once the ball is halfway to home plate. Given the speed of the pitch and the limitations of our biology, it seems like a miracle that anybody ever hits the ball at all.

Still, Albert Pujols and his All-Star peers see—and crush—95-mph fastballs for a living. So why are they transmogrified into Little Leaguers when faced with 68-mph softballs? It's because the only way to hit a ball traveling at high speed is to be able to see into the future, and when a baseball player faces a softball pitcher, he is stripped of his crystal ball.

———————

Nearly forty years ago, before Janet Starkes became one of the most influential sports expertise researchers in the world, she was a 5'2" point guard who spent one summer with the Canadian national team. Her lasting influence on sports, though, would come off the court, from the work she started as a graduate student at the University of Waterloo. Her research was to try to figure out why good athletes are, well, good.

Tests of innate physical "hardware"—qualities that an athlete is apparently born with, like simple reaction time—had done astonishingly little to help explain expert performance in sports. The reaction times of elite athletes always hovered around one fifth of a second, the same as the reaction times when random people were tested.

So Starkes looked elsewhere. She had heard of research on air traffic controllers that used "signal detection tests" to gauge how quickly an expert controller can sift through visual information to determine the presence or absence of critical signals. And she decided that conducting studies like these, of perceptual cognitive skills that are learned through practice, might prove fruitful. So, in 1975, as part of her graduate work at Waterloo, Starkes invented the modern sports "occlusion" test.

She gathered thousands of photographs of women's volleyball games and made slides of pictures where the volleyball was in the frame and others where the ball had just left the frame. In many photos, the orientation and action of players' bodies were nearly identical regardless of whether the ball was in the frame, since little had changed in the instant when the ball had just exited the picture.

Starkes then connected a scope to a slide projector and asked competitive volleyball players to look at the slides for a fraction of a second and decide whether the ball was or was not in the frame that had just flashed before their eyes. The brief glance was too quick for the viewer actually to see the ball, so the idea was to determine whether players were seeing the entire court and the body language of players in a different way from the average person that allowed them to figure out whether the ball was present.

The results of the first occlusion tests astounded Starkes. Unlike in the results of reaction time tests, the difference between top volleyball players and novices was enormous. For the elite players, a fraction of a second glance was all they needed to determine whether the ball was present. And the better the player, the more quickly she could extract pertinent information from each slide.

In one instance, Starkes tested members of the Canadian national volleyball team, which at the time included one of the best setters in the world. The setter was able to deduce whether the volleyball was present in a picture that was flashed before her eyes for sixteen thousandths of a second. "That's a very difficult task," Starkes told me. "For people who don't know volleyball, in sixteen milliseconds all they see is a flash of light."

Not only did the world-class setter detect the presence or absence of the ball in sixteen milliseconds, she gleaned enough visual information to know when and where the picture was taken. "After each slide she would say 'yes' or 'no,' whether the ball was there," Starkes says, "and then sometimes she would say, 'That was the Sherbrooke team after they got their new uniforms, so the picture must have been taken at such and such a time.'" One woman's blink of light was another woman's fully formed narrative. It was a strong clue that one key difference between expert and novice athletes was in the way they had learned to perceive the game, rather than the raw ability to react quickly.

Shortly after she received her Ph.D., Starkes joined the faculty at McMaster University and continued her occlusion work with the Canadian national field hockey team. At the time, the coaching orthodoxy in field hockey favored the idea that innate reflexes were of primary importance. Conversely, the idea that learned, perceptual skills were a hallmark of expert performance was, as Starkes put it, "heretical."

In 1979, when Starkes began helping the Canadian national field hockey team gear up for the 1980 Olympics, she was dismayed to find that the national coaches were relying on outdated ideas to choose and arrange the team. "They thought everybody saw the field the same way," she says. "They were using simple reaction time tests for selection, and they thought it would be a good determinant of who would be the best goalies or strikers. I was astounded that they had no idea that reaction time might not be predictive of anything."

Starkes, of course, knew better. In her occlusion tests of field hockey players, she found just what she had found in volleyball players, and more. Not only were elite field hockey players able to tell faster than the blink of an eye whether a ball was in the frame, they could accurately reconstruct the playing field after just a fleeting glance. This held true from basketball to soccer. It was as if every elite athlete miraculously had a photographic memory when it came to her sport. The question, then, is how important these perceptual abilities are to top athletes and whether they are the result of genetic gifts.

There is no better place to look for an answer than in a type of competition where the action is slow, deliberate, and devoid of the constraints of muscle and sinew.

In the early 1940s, Dutch chess master and psychologist Adriaan de Groot began drilling for the core of chess expertise. De Groot would test chess players of various skill levels and attempt to dissect what made a grandmaster better than an average professional, and the average professional far superior to a club player.

The common wisdom of the time was that highly skilled chess players thought further ahead in the game than did less skilled players. This is true when skilled players are compared with complete novices. But when de Groot asked both grandmasters and merely strong players to narrate their decision making in the face of an unfamiliar game situation, he found that players of disparate skill levels mulled over the same number of pieces and proposed essentially the same array of possible moves. Why then, he wondered, do the grandmasters end up making *better* moves?

De Groot assembled a panel of four chess players as representatives of their varying skill echelons: a grandmaster and world champion; a master; a city champion; and an average club player.

De Groot enlisted another master to come up with different chess arrangements taken from obscure games, and then did something very similar to what Starkes would do with athletes thirty years later: he flashed the chessboards in front of the players for a matter of seconds and then asked them to reconstruct the scenario on a blank board. What emerged were differences between the skill levels, particularly the two masters and the two nonmasters, "so large and unambiguous that they hardly need further support," de Groot wrote.

In four of the trials, the grandmaster re-created an entire board after viewing it for three seconds. The master was able to accomplish the same feat twice. Neither of the lesser players was able to reproduce any boards with complete accuracy. Overall, the grandmaster and master accurately replaced more than 90 percent of the pieces in the trials, while the city champion managed around 70 percent, and the club player only about 50 percent. In five seconds, the grandmaster understood more of the game situation than the club player did in fifteen minutes. In these tests, de Groot wrote, "it is evident that experience *is* the foundation of the superior achievements of the masters." But it would be three decades before confirmation would come that what de Groot saw was indeed an acquired skill, and not the product of innately miraculous memory.

In a seminal study published in 1973, Carnegie Mellon University psychologists William G. Chase and Herbert A. Simon—a future Nobel Prize winner—repeated the de Groot experiment, and added a twist: they tested the players' recall for chessboards that contained random arrangements of pieces that could never occur in a game. When the players were given five seconds to study the random assortments and then asked to re-create them, the recall advantages of the masters disappeared. Suddenly, their memories were just like those of average players.

In order to explain what they saw, Chase and Simon proposed a "chunking theory" of expertise, a pivotal idea in the study of games like chess, but also in sports, that helps explain what Janet Starkes found in her work with field hockey and volleyball players.

Chess masters and elite athletes alike "chunk" information on the board or the field. In other words, rather than grappling with a large number of individual pieces, experts unconsciously group information into a smaller number of meaningful chunks based on patterns that they have seen before. Whereas the average club player in de Groot's study was scanning and attempting to remember the arrangement of twenty individual chess pieces, the grandmaster needed to remember only a few chunks of several pieces each, because the relationships between the pieces had great meaning for him.*

A grandmaster is fluent in the language of chess and has a mental database of millions of arrangements of pieces that are broken down into at least 300,000 meaningful chunks, which are in turn grouped into mental "templates," large arrangements of pieces (or players, in the case of athletes) within which some pieces can be moved around without rendering the entire arrangement unrecognizable. Where the novice is overwhelmed by new information and randomness, the master sees familiar order and structure that allows him to home in on information that is critical for the decision at hand. "What was once accomplished by slow, conscious deductive reasoning is now arrived at by fast, unconscious perceptual processing," Chase and Simon wrote. "It is no mistake of language for the chess master to say that he 'sees' the right move."

Studies that track the eye movements of experienced performers, whether chess players, pianists, surgeons, or athletes, have found that

---

*We all use forms of chunking every day. Consider language: if I give you a twenty-word sentence to remember, you will have a much easier time repeating it than if I give you twenty random words that have no meaningful relationship to one another.

as experts gain experience they are quicker to sift through visual information and separate the wheat from the chaff. Experts swiftly move their attention away from irrelevant input and cut to the data that is most important to determining their next move. While novices dwell on individual pieces or players, experts focus more attention on the spaces between pieces or players that are relevant to the unifying relationship of parts in the whole.

Most important in sports, perceiving order allows elite athletes to extract critical information from the arrangement of players or from subtle changes in an opponent's body movements in order to make unconscious predictions about what will happen next.

Bruce Abernethy was an undergraduate at the University of Queensland in the late 1970s and an avid cricket player when he began to expand on Janet Starkes's occlusion methods. Abernethy started out using Super 8mm film to capture video of cricket bowlers. He would show batters the video but cut it off before the throw and have them attempt to predict where the ball was headed. Unsurprisingly, expert players were better at predicting the path of the ball than novice players.

In the decades since, Abernethy, now associate dean for research at Queensland, has become exceedingly sophisticated at using occlusion tests to illuminate the basis of perceptual expertise in sports. Abernethy has moved his studies from the video screen to the field and the court. He has equipped tennis players with goggles that go opaque just as an opponent is about to strike the ball, and he has outfitted cricket batters with contact lenses with varied levels of blurriness.

The theme of Abernethy's findings is that elite athletes need less time and less visual information to know what will happen in the future, and, without knowing it, they zero in on critical visual information, just like expert chess players. Elite athletes chunk information about bodies and player arrangements the way that grandmasters do with rooks and bishops. "We've tested expert batters in cricket where all they see is the ball, the hand and wrist, and down to the elbow, and they still do better than random chance," Abernethy says. "It looks bizarre, but there's significant information between the hand and arm where experts get cues for making judgments."

Top tennis players, Abernethy found, could discern from the minuscule pre-serve shifts of an opponent's torso whether a shot was going to their forehand or backhand, whereas average players had to wait to see the motion of the racket, costing invaluable response time. (In badminton, if Abernethy hides the racket and entire forearm, it transforms elite players back into near novices, an indication that information from the lower arm is critical in that sport.)

Pro boxers have a similar skill. A Muhammad Ali jab took a mere forty milliseconds to arrive at the face of a victim standing a foot and a half away. Without anticipation based on body movements, Ali's opponents would have been beaten down in round one, hit flush by every punch. (Ali's skill at disguising the trajectory of a punch, and thus confounding the opponent's anticipation, often meant they were finished a few rounds later anyway.)

Even skills that appear to be purely instinctive—jumping to rebound a basketball after a missed shot—are grounded in learned perceptual expertise and a database of knowledge on how subtle shifts of a shooter's body will alter the trajectory of the ball. It's a database that can be built only through rigorous practice.*

Without that database, every athlete is a chess master facing a random board, or Albert Pujols facing Jennie Finch, stripped of the information that allows him to predict the future.† Since Pujols had no mental database of Finch's body movements, her pitch tendencies, or

---

*Pro cricket teams have been moving away from using bowling machines, because they don't train the body recognition skills that hitters need for anticipation.
†According to analysis by hitting coach Perry Husband of all 500,000 pitches from one full MLB season, on pitches that were directly down the middle major leaguers hit .462 when the count was two balls and zero strikes, and .362 when the count was zero balls and two strikes—a 100-point difference based solely on count information that helped hitters to anticipate the next pitch.

even the spin of a softball to predict what might be coming, he was always left reacting at the last moment. And Pujols's simple reaction speed is downright quotidian.

When scientists at Washington University in St. Louis tested him, Pujols, the greatest hitter of an era, was in the sixty-sixth percentile for simple reaction time compared with a random sample of college students.

No one is born with the anticipatory skills required of an elite athlete. When Abernethy studied the eye movement patterns of elite and novice badminton players, he saw that the novices were already looking at the correct area of the opponent's body, they just did not have the cognitive database needed to extract information from it. "If they did," Abernethy says, "it would be a hell of a lot easier to coach them to become an expert. You could just say, 'Look at the arm. Or for a baseball batter the real advice wouldn't be 'keep your eye on the ball,' it would be 'watch the shoulder.' But actually, if you tell them that, it makes good players worse."

As an individual practices a skill, whether it be hitting, throwing, or learning to drive a car, the mental processes involved in executing the skill move from the higher conscious areas of the brain in the frontal lobe, back to more primitive areas that control automated processes, or skills that you can execute "without thinking."

In sports, brain automation is hyperspecific to the practiced skill, so specific that brain-imaging studies of athletes who train in a particular task show that activity in the frontal lobe is turned down only when they do that exact task. When runners are put on bicycles or arm bikes (where the pedals are moved with hands instead of feet) their frontal lobe activity increases compared with when they are running, even though cycling or arm cycling wouldn't seem to require much conscious thought. The physical activity that one trains in is very specifically automated in the brain. To return to Abernethy's point, "thinking" about an action is the sign of a novice in sports, or a key to transforming an expert back into an amateur. (University of Chicago psychologist Sian Beilock has shown that a golfer can overcome pressure-induced choking in putting—paralysis by analysis, she calls it—by singing to himself, and thus preoccupying the higher conscious areas of the brain.)

Chunking and automation travel together on the march toward expertise. It is only by recognizing body cues and patterns with the rapidity of an unconscious process that Albert Pujols can determine whether he should swing at a ball when it has barely left the pitcher's hand. The same goes for quarterback Peyton Manning. He cannot stop in the face of blitzing linebackers and consciously sort through the defensive alignments and patterns he learned in hours and years of practicing and studying game film. He has seconds to scan the field and throw. He is a grandmaster playing speed chess, only with linebackers and safeties in place of knights and pawns. (At the same time, NFL defensive coordinators are shuffling their players in an attempt to present Manning with a chessboard that looks misleading or random.)

The result of expertise study, from de Groot to Abernethy, can be summarized in a single phrase that played like a broken record in my interviews with psychologists who research expertise: "It's software, not hardware." That is, the perceptual sports skills that separate experts from dilettantes are learned, or downloaded (like software), via practice. They don't come standard as part of the human machine. That fact helped spawn the most well-known theory in modern sports expertise, and one that has no place for genes.

It started with musicians.

For a 1993 study, three psychologists turned to the Music Academy of West Berlin, which had a global reputation for producing world-class violinists.

The academy professors helped the psychologists identify ten of the "best" violin students, those who could become international soloists; ten students who were "good" and could make a living in a symphony orchestra; and ten lesser students they categorized as "music teachers," because that would be their likely career path.

The psychologists conducted detailed interviews with all thirty academy students, and certain similarities emerged. All of the musicians from all three groups had started taking systematic lessons at around eight years old, and all had decided to become musicians around fifteen. And, despite their skill differences, the violinists from all three groups dedicated a whopping 50.6 hours each week to their music skills, whether taking music theory classes, listening to music, or practicing and performing.

Then a major difference surfaced. The amount of time that the violinists in the top two groups spent practicing on their own: 24.3 hours each week, compared with 9.3 for the bottom group. Perhaps not surprisingly, then, the musicians rated solitary practice as the most important aspect of their training, albeit a much more taxing one than activities like group practice or playing for fun. Everything in the lives of the violinists in the top two groups seemed to orbit around training and recovery from training. They slept 60 hours each week, compared with 54.6 for the bottom group. But even the hours spent practicing alone didn't differentiate the top two groups.

So the psychologists asked the violinists to make retrospective estimates of how much they had practiced since the day they began playing. The top violinists had begun ramping up their practice hours more quickly after they first took up the instrument. By age twelve, the best violinists had a head start of about 1,000 hours on the future teachers. And even though the top two groups were spending identical amounts of time on their craft at the academy, the future international soloists had accumulated, on average, 7,410 hours of solitary practice by age eighteen, compared with 5,301 hours for the "good" group, and 3,420 hours for the future teachers. "Hence," the psychologists wrote, "there is complete correspondence between the skill level of the groups and their average accumulation of practice time alone with the violin." In essence, they concluded that what might have been construed as innate musical talent was actually years of accumulated practice.

Remarkably, the psychologists found that expert pianists had, on average, accumulated a similar number of practice hours as the top violinists, as if there were some universal rule of expertise. The researchers used the weekly practice estimates to suggest that expert musicians, regardless of the instrument, accumulate 10,000 hours of practice by age twenty, and that skilled performers engage in greater quantities of "deliberate practice," the kind of effortful exercises that strain the capacity of the trainee. The kind of practice that is often done in solitude.

In the now-famous paper—"The Role of Deliberate Practice in the Acquisition of Expert Performance"—the authors extended their conclusions to sports, citing Janet Starkes's occlusion tests that showed learned perceptual expertise is more important than raw reaction skills. Accumulated hours of practice, they suggested, were masquerading as innate talent in both music and sports.

The lead author of the paper, psychologist K. Anders Ericsson, now at Florida State, came to be viewed as the father of the "10,000 hours" to expertise rule—though he himself never called it a "rule"—or the "deliberate practice framework," as it is often known among those who study skill acquisition.

Ericsson is regarded as an expert on experts. He and other proponents of the framework went on to suggest that accumulated practice is the real wizard behind the curtain of innate talent in fields from sprinting to surgery.

As genetic science became more prominent, Ericsson worked genes into his writing. In a 2009 paper, "Toward a Science of Exceptional Achievement," Ericsson and his coauthors write that the genes necessary to be a pro athlete (or a pro anything, really) "are contained within all healthy individuals' DNA." In that view, experts are differentiated by their practice histories, not their genes. The media interpretation of Ericsson's work has often been to say that 10,000 hours is both necessary and sufficient to make anyone an expert in anything. No one, the idea goes, achieves expertise with less, and everyone achieves expertise with that amount.

On the backs of several bestselling books and reams of articles, the 10,000-hours rule (alternately known as the ten-year rule) has become embedded in the world of athlete development and an impetus for starting children early in hard training.

In some cases, popular writers describing Ericsson's work have allowed for individual genetic differences in addition to differences born of practice, while others have taken a rigid view of the 10,000-hours rule as absolute, with no room for genetic gifts. During the reporting of this book, I saw the 10,000 hours referenced as the recipe for success in arenas as disparate as an interview given by a U.S. Olympic Committee scientist and the annual letter from a hedge fund to its investors explaining the fund's tenets of success.

I even became acquainted with a golfer who is putting the rule to a very personal test.

2

A Tale of Two High Jumpers

(Or: 10,000 Hours Plus or Minus 10,000 Hours)

On June 27, 2009, his thirtieth birthday, Dan McLaughlin resolved to do something special: quit his job as a commercial photographer in Portland, Oregon, and become a professional golfer. His golf experience over the previous three decades consisted of two childhood trips to a driving range with his older brother. Save for some youth tennis and a season of cross-country running in high school, McLaughlin hadn't been a competitive athlete. But something had to change.

After completing his journalism degree at the University of Georgia in 2003, he took pictures for newspapers for two years, and then worked in various forms of advertising and product photography. After six years at a desk job that centered on snapping photos of dental equipment, McLaughlin needed a venture more suited to his taste for challenge.

At first, he thought it might be grad school. So he saved enough money to start an MBA program in finance. But it took only the first day's class at Portland State, on how to operate Microsoft Excel spreadsheets, for McLaughlin to realize that an MBA was not the change of course he craved. He mulled over becoming a physician's assistant, or an architect, but decided that the new path had to be drastic.

McLaughlin had always had a bit of the extreme in him. His idea of a winter vacation in 2006 was a trip to Fiji during the nation's military coup. And yet, in many ways, McLaughlin is the Everyman. He's 5'9", 150 pounds, and "not particularly physically gifted," in his own words. "I'm kind of just a very average-type person," he says. That's what he's counting on.

McLaughlin was inspired by what he read of Ericsson's work in the bestsellers *Talent Is Overrated*, by Geoff Colvin, and *Outliers*, by Malcolm Gladwell. He read about the 10,000-hours rule, the "magic number of greatness," as it is called in *Outliers*, and about the idea that skills that appear to be predicated on innate gifts are often nothing more than the manifestations of thousands of hours of practice.

And so it was that on April 5, 2010, McLaughlin logged his first two hours of deliberate practice toward his ultimate goal of going pro and making the PGA Tour. His plan is to log every single hour along the path to 10,000, and to show that "there's no difference between experts and me, or other people, not just in golf, but in any field. If I were over six feet tall, that might not speak to most people, but I'm a normal guy."

McLaughlin is not approaching his journey—he had logged 3,685 hours by the end of 2012—as a publicity stunt, but as a scientific experiment. He enlisted a PGA-certified instructor and consults with Ericsson for advice on his strategy. McLaughlin is committed to counting only those hours of practice that truly qualify as deliberate according to Ericsson's definition.

"According to the tenets of deliberate practice, you have to be cognitively engaged," McLaughlin explains. Just going to the driving range and swatting balls for a few hours without an eye toward improvement and error correction doesn't cut it. So, six days a week, McLaughlin puts in six hours of deliberate practice, a workday that consumes eight hours because he takes frequent breaks to think about what he did well and what can be improved—like closing the club face on impact—and because it is exhausting to maintain strict focus for hours on end.

McLaughlin is building his golf game from the ground up. When I first spoke with him, 1,776 hours into his journey, he had yet to wield a driver. "I'm only up to an eight-iron," he said, "so my game is all within 140 yards of the hole." On the occasions when McLaughlin decides to play something resembling a round with his eight-iron, he places three balls at varying distances from the cup and plays all three at once. "That way," he says, "I can get twenty-seven holes of play in on only nine holes." At his current pace, McLaughlin will reach 10,000 hours late in 2016. (And he isn't even counting the hours he spends lifting weights, reading golf theory, or working with a nutritionist.) McLaughlin fully expects to be a professional when he reaches the magic number. "There are no guarantees," he says. "I could get in a car wreck and die tomorrow. But my ultimate goal is to make the PGA Tour."

"No matter what happens," he continues, "I will consider it a success. I love the game more every day, and I gave a presentation at a conference at Florida State, where I had breakfast, lunch, and dinner with Dr. Ericsson. . . . He said this is useful for him to see how things progress, even though it's just one person. He said he's never done this long a study on someone, tracking their deliberate practice."

No one has ever done such a study. All of the data in support of the 10,000-hours rule have been what scientists call "cross-sectional" and "retrospective." That is, the researchers look at subjects who have already attained a certain skill level and ask them to reconstruct their history of practice hours. In the case of the original 10,000-hours study, the subjects were musicians who had already gained admission to a world-famous academy, so most of humanity had long since been screened out. A study that is restricted to only prescreened performers is hopelessly biased *against* discovering evidence of innate talent. A "longitudinal" study, on the other hand, is a much higher standard of experimentation that follows subjects as they accumulate those hours in order to watch how their skills progress. It's easy to understand why longitudinal research of the 10,000-hours rule is difficult: imagine the challenge of recruiting a group of Dan McLaughlins for a study—all willing to spend years practicing a skill they've never tried—much less tracking them assiduously.

There is, however, a way to track the acquisition of skill expertise without at least some of the problems of subjective human recall.

Chess players are rated according to Elo points, named for Arpad Elo, a physicist who created the ranking system. An average chess player has around 1,200 Elo points. A master, the bare minimum level to make a living playing chess, has between 2,200 and 2,400 points. An international master has 2,400 to 2,500, and a grandmaster has more than 2,500 Elo points. Because Elo points are accumulated as a player improves, the rating system provides an objective accounting of a player's historical skill progression.

In 2007, psychologists Guillermo Campitelli, of the Universidad Abierta Interamericana in Buenos Aires, and Fernand Gobet, director of the Centre for the Study of Expertise at Brunel University in West London, recruited 104 competitive chess players of varying skill levels for a study of chess expertise. Campitelli had coached future grandmasters, and Gobet, who logged eight to ten hours a day of chess practice in his youth, had been an international master and the second-ranked player in Switzerland.

Campitelli and Gobet found that 10,000 hours was not far off in terms of the amount of practice required to attain master status, or 2,200 Elo points, and to make it as a pro. The average time to master level in the study was actually about 11,000 hours—11,053 hours to be exact—so more than in Ericsson's violin study. More informative than the average number of practice hours required to attain master status, however, was the *range* of hours.

One player in the study reached master level in just 3,000 hours of practice, while another player needed 23,000 hours. If one year generally equates to 1,000 hours of deliberate practice, then that's a difference of two decades of practice to reach the same plane of expertise. "That was the most striking part of our results," Gobet says. "That basically some people need to practice eight times more to reach the same level as someone else. And some people do that and still have not reached the same level."* Several players in the study who started early in childhood had logged more than 25,000 hours of chess practice and study and had yet to achieve basic master status.

While the average time to master level was 11,000 hours, one man's 3,000-hours rule was another man's 25,000-and-counting-hours rule. The renowned 10,000-hours violin study only reports the *average* number of hours of practice. It does not report the *range* of hours required for the attainment of expertise, so it is impossible to tell whether any individual in the study actually became an elite violinist in 10,000 hours, or whether that was just an average of disparate individual differences.

On a panel at the 2012 American College of Sports Medicine conference, Ericsson noted that the now world-famous data were collected in a small number of subjects and are not entirely reliable in terms of counting practice hours. "Obviously, we were only collecting data on ten individuals," Ericsson said. "And [the violinists did] some of the retrospective estimates several times, and there was no perfect agreement." That is, the violinists were inconsistent in multiple accounts of how much they had practiced. Even so, Ericsson said, the variation among just the ten most elite violinists—the 10,000-hours group—was still "certainly more than 500 hours." (Ericsson himself, it should be noted, never used the term "10,000-hours rule." In a 2012 paper in the *British Journal of Sports Medicine,* he ascribed the phrase's popularity to a chapter title in Malcolm Gladwell's *Outliers,* which, he wrote, "misconstrued" the conclusions of the violin study.)

When I asked Dan McLaughlin whether he had any concern that he might, like some of the chess players, be a 20,000-hours guy as opposed to a 10,000-hours guy, he said that he considered the journey a victory in itself. "When it comes down to D-Day and it's my ten-thousandth hour,"

---

*Another striking result was that chess pros were twice as likely as non–chess pros to be left-handed.

McLaughlin said, "it'll be interesting to see whether I'm still shooting seventy-five, or I missed Q-School [the PGA Tour's qualifying school] by one stroke, or if I'm on the Tour. I think you could probably master something in anywhere from 7,000 to 40,000 hours, but this is kind of a good way to keep track of progress." Somehow, the 7,000-to-40,000-hours rule just doesn't have the same ring to it.

For the chess players, differences in progress showed up right away. "If you look at those players who go on to be masters and those who remain below that level," Gobet says, "some of them have the same practice the first three years, but there were already large differences in performance. Perhaps if there are very small individual differences [in talent] at the beginning, they make a huge effect. We assume it takes about ten seconds to learn a chunk, and we have estimated that it takes about 300,000 chunks to become a grandmaster. If one person learns each chunk in nine seconds and the other person eleven seconds, those small differences are going to be amplified."

It's a sort of butterfly effect of expertise. If two practitioners start with slightly different initial conditions, according to Gobet, it can lead to dramatically different outcomes, or at least to drastically different amounts of practice that will be required for similar outcomes.

On the morning of August 22, 2004, Stefan Holm was staying calm the way he always did before a competition, by losing himself in a book. This time, it was *Olympics in Athens 1896: The Invention of the Modern Olympic Games* by Michael Llewellyn Smith. When Holm, a Swedish high jumper, traveled for competitions, he liked to choose books that were relevant to the locale he was visiting. And this book was particularly apropos, as he would be competing in the *Olympiakó Stádio* in Athens in a few hours in the 2004 Olympic final.

As always, Holm made sure that he forced every omen into auspicious alignment. Even if he wanted to stop reading his book at page 225, he would make sure to read to at least page 240, because when the bar was raised to 225 centimeters (7'5") during the competition, he did not want that number associated with stopping in his mind.

In order to avoid the mental strain of small decisions, Holm's morning followed a practiced pattern: first, corn flakes and orange juice for breakfast; then, one hour before leaving for the track, he laid the blue and yellow competition clothes bearing the symbol of the Swedish crown on the bed, followed by a shower, shampoo—always twice, for no reason he could explain—and a shave. He packed his bag in the same order every time. He wore the same black underwear he always did for competition. He put the right sock on before the left, and his jumping shoes in the reverse order, the left before the right.

At the track that evening, Holm's life came down to one final attempt at 7'8". He missed on his first two jumps. A third miss would be the end. As he did before every jump, he whisked his hands backward over his shorn hair, twice, wiped his eyes, tugged the chest of his jersey, and then cleared the sweat from his brow. He took a few baby steps toward the bar, and then broke into a full sprint. He launched himself into the air, and sailed right over. After that, he cleared 7'9" to win the Olympic gold medal. It was a fitting climax to a story that began with the kind of youthful obsession that is capable of producing genius.

Inspired by the Moscow Olympics, Holm took his first jumps with his neighbor Magnus over the sofa when he was just four years old, in 1980, an adventure that ended when Magnus broke his arm. But the duo was undeterred.

When Holm was six, Magnus's father built a high jump pit for the boys from pillows and an old mattress and placed it in the backyard. Two years later, in 1984, when Holm was eight, he saw a competition featuring Patrik Sjöberg, the brash Swedish jumper with the cascading golden tresses who would go on to set the world record. All across Sweden, hordes of mini-Sjöbergs began scissor-kicking and Fosbury-flopping over their parents' couches. The young Holm often beckoned his father's attention with delighted squeals of "Look! I'm Patrik Sjöberg!" before bounding over the couch.

Holm started school around that time, an endeavor that excited him primarily because the school had a high-jump pit. He spent many a lunch hour, with Magnus, enacting a fantasy version of the Olympic high-jump competition, occasionally showing up tardy for class.

On the day of the Athens final, Magnus was there in the stands, and so was Johnny Holm, Stefan's father and lifelong coach. In his youth, Johnny Holm had been a catlike goalkeeper in Sweden's fourth division, and could have progressed toward the professional ranks, but he chose to stay close to home and to his job as a welder. From the time Stefan Holm was a teenager, he could sense from his father's stories that Johnny regretted never having taken the chance to become a professional athlete. His father did not say it outright, but Holm could tell from how eager Johnny was to help his son throw himself fully into high jump. Both Holm and his father became obsessed with the sport.

In 1987, as if sent by the jumping gods to aid Stefan Holm in his quest, a professional-grade indoor track-and-field facility called Våxnäshallen opened in western Sweden, just a few minutes' drive from his tiny hometown of Forshaga. It gave Holm, at the age of eleven, what would become his year-round, career-long, world-class training venue.

At fourteen, Holm cleared six feet, an age-group record in his area in the west of Sweden, though he was defeated at a handful of competitions that season. At fifteen, he won the Swedish youth championships and traveled with his father to Gothenburg to meet Patrik Sjöberg's coach, Viljo Nousiainen. The meeting sparked an enduring friendship between the elder Holm and Nousiainen, and Johnny Holm began to adapt some of Nousiainen's training methods for his teenage son. The boy who had idolized the great Patrik Sjöberg was suddenly being groomed to become him. But there was an obvious difference. Sjöberg was 6'7", while every local newspaper article about Holm's accomplishments noted his diminutive stature. As an adult, Holm would top out at 5'11", downright Lilliputian for a high jumper. In a sport that requires raising one's center of mass as high as possible, starting with a high center of mass is an enormous advantage.

As a teenager, Holm developed the high jumper's equivalent of stage fright: when the bar was raised to a height above his head, he would take his normal approach, but rather than jumping he simply ran under the bar and onto the landing mat. In several competitions in his teens, Holm did that three straight times at a given height, which meant he was out of the competition. Instead of giving up, Holm redoubled his work, quitting soccer and dedicating himself solely to high jump. At sixteen, he lost only a single competition—a wound he would remember and avenge with his undefeated 2004 season—and immersed himself in what he later called a "twenty-year love affair with the high jump." (For much of those two decades, it was an exclusive love affair that left Holm little time for girl-friends.) As Holm himself acknowledges, it would be a fair bet that he has taken more high jumps than any human being who has ever lived.

By seventeen, Holm was good enough to face his hero Sjöberg in competition. Sjöberg won handily, but Holm wondered whether he could one day top the Swedish icon if he kept at it. At nineteen, Holm started a weight-lifting regimen—concentrated on his left leg, of course—that would get progressively more intense over a decade to the point where he could put 310 pounds, double his weight, on his shoulders and squat so low that his butt nearly grazed the ground, before popping back up.

To compensate for his stature, Holm perfected a sprinting approach where he hit a top speed around nineteen miles per hour, likely faster than any other jumper in the world. To accommodate that speed, he had to start taking off from farther and farther away from the bar. Holm was flying faster, farther, and higher every year, rocketing at the bar and curling his body around it so tightly that if his heels had a secret they could whisper it into his ear when he was in full arch. Starting in 1987, Holm improved a few centimeters every year, without fail. In a task that seems so "you either got it, or you don't," Holm was transforming himself into the ultimate "got it."

In 1998, Holm won the first of eleven consecutive Swedish national championships. Three years later, he finished just off the Olympic medal stand, taking fourth in Sydney. That was not good enough.

Holm had been living at home and taking college classes on and off. At twenty-five, he dropped out of school and moved out of his parents' house into an apartment that was just down the road from the Våxnäshallen facility, in Karlstad, a town of sixty thousand that sits on the north coast of the largest lake in Sweden. From then on, Holm trained twelve sessions per week. His workday started at ten A.M. with two hours of weights, box jumps, or hurdles—he and his father designed hurdles that could be raised to five and a half feet. Then a break for lunch, and another session in the late afternoon that might consist of thirty high jumps at full competition speed. Thirty, that is, if all went according to plan. Holm could not go home on a miss, nor would he lower the bar to facilitate a clearance, so practice went until he made it over whatever height he was confronting. By the time Athens rolled around, Johnny Holm had watched his son take so many jumps that he could tell whether Stefan would clear the bar when he was still four steps from liftoff.

Without a running start, Holm's standing vertical jump hovered around twenty-eight inches, which is perfectly pedestrian for an athlete. But his blazing fast approach allowed him to slam down on his Achilles tendon, which would then act like a rebounding spring to propel him over the bar. When scientists examined Holm, they determined that his left Achilles tendon had hardened so much from his workout regimen that a force of 1.8 tons was needed to stretch it a single centimeter, about four times the stiffness of an average man's Achilles, making it an unusually powerful launching mechanism.

In 2005, a year after he won the Olympic title, Holm earned a qualification of the perfect human projectile: he cleared 7'10.5", equaling the record for the highest high-jump differential between the bar and the jumper's own height.

Late in the day that I met him at the snow-covered train station in Karlstad, Holm took me to Våxnäshallen, the facility that "was my home for twenty years," he said. On one side of the track, near a weight-lifting area, is a locked box that contains Holm's custom-made hurdles. To save himself from himself, Holm has given away the key. He still visits to high jump once or twice a week, though, and his father trains young jumpers at the facility.

Holm's son Melwin has begun to tag along. (Melwin is not a Swedish name. Holm and his wife liked "Melvin," and Holm wanted "win" somewhere in the boy's name.) One day in 2007, when Melwin was two and Johnny Holm was babysitting, Stefan returned home and found his infant son flopping backward in diapers over a high jump constructed of Lego Duplo bricks. "He cleared thirty centimeters," Holm says, with a straight face.

At Våxnäshallen, a few kids approach Holm for autographs. (Since his athletic retirement, Holm has become famous for winning Swedish television quiz shows. He has a steel trap of a memory and can recall exact heights from competition jumps twenty years in the past.) For the most part, Holm is left alone to watch a group of seven- and eight-year-old children try high jumping. Some of the kids jump off the wrong foot. Others go off both feet. As the children flop onto the mat one by one, Holm points to those who have a feel for how their body should move in the air. Holm whispers to me, noting the children who he thinks have potential. When I ask whether he could teach any one of them to be an Olympic champion, he says: "There are some things you can't teach, the sort of feel for jumping. I was never into training the technical things. The [back] arch was just always there."

As we leave the facility and make our way back toward the train station, we pass a bookstore. "Come here," Holm beckons, pointing through the window of the store at a white book bearing a hand that is painted blue and making a victory sign. As I press my face to the glass, I see that it's the Swedish translation of Malcolm Gladwell's *Outliers*.

"You see this? Read this," Holm says. "There were jumpers who beat me when I was young. You wouldn't have said I would be Olympic champion. It's all about your ten thousand hours."

In 2007, Holm entered the World Championships in Osaka, Japan, as the favorite. And, despite the fact that there has never been a more assiduous student of high jump, Holm was faced with a competitor who he barely knew: Donald Thomas, a jumper from the Bahamas. Thomas had just begun high jumping. As Thomas's cousin, a college track coach, put it, "He still doesn't know that a track goes around in a circle."

The previous year, on January 19, 2006, Thomas was sitting in the cafeteria at Lindenwood University in Saint Charles, Missouri, boasting about his slam-dunking prowess with a few guys from the track team. Carlos Mattis, Lindenwood's top high jumper, had enough of Thomas's lip and bet him that he could not clear 6'6" in a high jump competition.

Thomas decided to put his hops where his mouth was. He went home and grabbed a pair of sneakers and returned to the Lindenwood field house where a smirking Mattis had already set the bar at 6'6". Mattis stepped back and waited for the big talker to fall to earth. And Thomas did, but the bar did not come with him. To Mattis's amazement, Thomas cleared it easily. So Mattis pushed the bar up to 6'8". Thomas cleared it. Seven feet. Without a semblance of graceful high-jump technique—Thomas hardly arched his back and his legs flailed in the air like the streamers trailing a kite—he cleared it.

Mattis rushed Thomas over to the office where head track coach Lane Lohr was organizing his roster for the upcoming Eastern Illinois University Mega Meet and told the coach that he had a seven-foot high jumper. "The coach said there's no way I could do that. He didn't believe it," Thomas recalls. "But Carlos was like, 'Yeah, he really did it.' So he asked if I wanted to go to a track meet on Saturday." Lohr picked up the phone and pleaded with the meet organizer to permit a late entry.

Two days later, in a black tank top and white Nike sneakers and shorts so baggy they blanketed the bar as he passed over it, Thomas cleared 6'8.25" on his first attempt, qualifying for the national championships.

Then he cleared 7'0.25" for a new Lindenwood University record. And then, on the seventh high jump attempt of his life, with rigid form akin to a man riding an invisible deck chair backward through the air, Thomas cleared 7'3.25", a Lantz Indoor Fieldhouse record. That's when Coach Lohr forced him to stop out of concern that he might hurt himself.

It would get better. Two months later, Thomas competed at the Commonwealth Games in Australia against some of the best professional jumpers in the world, wearing tennis shoes. He placed fourth in a world-class field, a result that actually confused him because he did not yet understand how tiebreakers work in high jump and thought that he was in third place until the results were announced.

Thomas's cousin Henry Rolle was the hurdles coach at Auburn University, and Thomas was swiftly offered a scholarship to Auburn on the condition that he agree to commit to actually start training for the high jump in 2007. So he did. Sort of.

Auburn assistant coach Jerry Clayton had coached Charles Austin, the 1996 Olympic high jump champion, and saw right away that he needed to develop Thomas slowly. "When he first got here, he didn't know how to warm up or stretch," Clayton said. And then there was the issue of practice. Thomas would step out of practice at Auburn's Beard-Eaves-Memorial Coliseum under the guise of going for a drink of water, and forty minutes later Clayton would find him outside shooting baskets. In Thomas's own words, he found high jump "kind of boring."

With a few months of light training, Clayton lessened Thomas's stutter-step, and though he couldn't get Thomas to put on the high jump shoes that every other elite competitor wears, he at least got him into pole vault shoes. In his first full season, Thomas cleared 7'7.75" to win the NCAA indoor high jump championship.

In August 2007, with a total of eight months of legitimate high-jump training to his name, Thomas donned his pole vault shoes and the gold and aquamarine uniform of his native Bahamas and traveled to Osaka,

Japan, for the World Championships. In non-Olympic years, the World Championships are the Super Bowl of track and field.

Thomas advanced easily to the final, as did Stefan Holm. When the men's high jump finalists were introduced, broadcasters announced a laser-focused Holm as the favorite. Thomas, looking cool in sunglasses beneath the bright lights illuminating the stadium, was described as "very much an unknown quantity."

Early in the competition, it appeared that Thomas would fold in his first world spotlight. While the rest of the jumpers took such lengthy approaches that they had to start on the running track, Thomas began on the infield, as if he were using the high jump equivalent of the short tees at a golf course. He stutter-stepped his way to a miss at 7'3"—each jumper gets three attempts at every height—lower than he jumped in that first meet at Eastern Illinois. Meanwhile, Holm was cruising, passing over 7'3", 7'5", 7'6.5", and 7'7.73" without a single miss, as his father watched through a video camera and pumped his fist in the stands.

But Thomas began to hit his form, managing to alternate makes and misses. He arrived at 7'8.5" along with a handful of other jumpers, including Holm.

For his first attempt, Holm stood with his eyes closed, envisioning himself floating over the bar. He approached, leapt, and barely grazed the bar. As it fell to the ground, he executed a frustrated backflip on the mat. Next, Yaroslav Rybakov, a 6'6" Russian, nudged the bar off the stand. Then came Thomas. He slowed down so drastically as he approached the bar that it seemed impossible that he could clear it. And yet, flailing his legs and with his back nearly straight, he passed 7'8.5" on his first attempt, putting his hand down behind him as if to break his fall because he was still uncomfortable with the sensation of falling backward. He rolled off the mat and gamboled across the track in celebration. But Holm was up again.

Another miss, just barely. Holm shook his palms in front of him as if beseeching the high jump gods. They didn't listen. On his final attempt, Holm clipped the bar with the back of his legs and fell to the mat with his head in his palms.

The guy in pole vault shoes who thinks high jump is "kind of boring" was crowned the 2007 world champion. On his winning jump, Thomas had raised his center of mass to 8'2". Had he any semblance of the back arch that every other pro jumper does, he would have shattered the world record.

Holm was polite in his remarks afterward, congratulating the new champion. Rybakov called Thomas's feat amazing, and noted that he himself had been practicing for an outdoor track-and-field world title for eighteen years and had yet to win one, compared with Thomas's eight months. But Johnny Holm, Stefan's coach and father, was so unnerved by Thomas's win that in a postevent interview he called him a "*jävla pajas*," literally "damn clown," essentially the Swedish equivalent of "buffoon." Johnny Holm said that Thomas's "flutter kick style" was a scandal for high jump, and suggested that the inelegance of his jumping was an affront to the sport and the men who had spent years training.

In 2008, the Japanese television station NHK asked Masaki Ishikawa, then a scientist at the Neuromuscular Research Center at the University of Jyväskylä in Finland, to examine Thomas. Ishikawa noted both Thomas's long legs relative to his height, and also that he was gifted with a giant's Achilles tendon. Whereas Holm's Achilles was a more normal-sized, incredibly stiff spring, Thomas's, at ten and a quarter inches, was uncharacteristically long for an athlete his height. The longer (and stiffer) the Achilles tendon, the more elastic energy it can store when compressed. All the better to rocket the owner into the air.

"The Achilles tendon is very important in jumping, and not just in humans," says Gary Hunter, exercise physiologist at the University of Alabama–Birmingham, and an author of studies on Achilles tendon lengths. "For example, the tendon in the kangaroo that's equivalent to our Achilles tendon is very, very long. That's why they can bounce around more economically than they can walk."

Hunter has found that a longer Achilles tendon allows an athlete to get more power from what's called the "stretch shortening cycle," basically the compression and subsequent decompression of the springlike tendon. The more power that is stored in the spring when it is compressed, the more you get when it's released. (A typical example is a standing vertical jump, in which the jumper bends down quickly, shortening the tendons and muscles, before jumping skyward.) When Hunter put subjects on a leg-press machine and dropped weights down on them, the longer the person's Achilles tendon the faster and harder he was able to fling the weights back in the opposite direction. "That's not exactly the same as a jump," Hunter says, "but it has a lot of similarities. And that's why people jump higher when they have a drop step or a few steps: they use the velocity of descent toward the ground to compress the tendon, just like a spring."

Tendon length is not significantly impacted by training, but rather is primarily a function of the distance between the calf muscle and heel bone, which are connected by the tendon. And while it appears that an individual can increase tendon stiffness by training, there is also growing evidence that stiffness is partly influenced by an individual's versions of genes involved in making collagen, a protein in the body that builds ligaments and tendons.

Neither Ishikawa nor Hunter would suggest that the sole secret to the jumping success of Holm and Thomas is in their Achilles tendons. But the tendons are one puzzle piece that helps explain how two athletes could arrive at essentially the same place, one after a twenty-year love affair with his craft, and the other with less than a year of serious practice after stumbling into it on a friendly bet. Interestingly, Thomas has not improved one centimeter in the six years since he entered the professional circuit. Thomas debuted on top and has not progressed. He seems to contradict the deliberate practice framework in all directions.

In fact, in absolutely every single study of sports expertise, there is a tremendous range of hours of practice logged by athletes who reach the same level, and very rarely do elite performers log 10,000 hours of sport-specific practice prior to reaching the top competitive plane, often competing in a number of other sports—and acquiring a range of other athletic skills—before zeroing in on one. A study of ultraendurance triathletes found that the better athletes had practiced far more on average but that there was a tenfold difference in practice hours among athletes who performed similarly.

Studies of athletes have tended to find that the top competitors require far less than 10,000 hours of deliberate practice to reach elite status. According to the scientific literature, the average sport-specific practice hours to reach the international levels in basketball, field hockey, and wrestling are closer to 4,000, 4,000, and 6,000, respectively. In a sample of Australian women competing in netball (sort of like basketball but without dribbling or backboards), arguably the best player in the world at the time, Vicki Wilson, had compiled only 600 hours of practice when she made the national team. A study of athletes on Australia's senior national teams found that 28 percent of them started their sport at an average age of seventeen, having previously tried on average three other sports, and debuted at the international level just four years later.

Even in this age of hyperspecialization in sports, some rare individuals become world-class athletes, and even world champions, in sports from running to rowing with less than a year or two of training. As with Gobet's chess players, in all sports and skills, the only real rule is that there is a tremendous natural range.

In 1908, Edward Thorndike, who would become known as the father of modern educational psychology, came up with a way to test whether nature or nurture dominated an individual's ability at a task. Thorndike was a leading proponent of the then-controversial idea that older adults—meaning, at the time, those over thirty-five—can continue to learn new skills. He figured that the way to distinguish nature from nurture was to give people the same amount of practice at a certain task and to see whether they became more or less alike. If their skill levels converged, Thorndike reasoned, then the impact of practice was overwhelming any innate individual differences. If they diverged, then nature was overpowering nurture.

In one experiment, Thorndike had adults practice multiplying three-digit numbers by three-digit numbers in their heads as quickly as they could. He was astounded by their improvement. "The fact that these mature and competent minds improved in the course of so short a training so much," Thorndike wrote, "is worthy of attention." After one hundred practice trials, many of the subjects cut their mental computation time in half. And every single subject improved. Just as in chess, language, music, and baseball, as practitioners improve at mental multiplication, they internalize patterns and systems of breaking problems into pieces that allow for increasingly rapid calculation.

But while Thorndike saw across-the-board improvement, he also noted what sociologists often call a "Matthew effect." The term derives from a passage in the biblical Gospel of Matthew:

For to all those who have, more will be given, and they will have an abundance; but from those who have nothing, even what they have will be taken away.

Thorndike saw that the subjects who did well at the start of the training also improved faster as the training progressed compared with the subjects who began more slowly. "As a matter of fact," Thorndike wrote, "in this experiment the larger individual differences *increase* with equal training, showing a positive correlation with high initial ability with ability to profit by training." The passage from the Bible doesn't quite capture Thorndike's results accurately because every subject improved, but the rich got relatively richer. Everyone learned, but the learning rates were consistently different.

When World War I erupted, Thorndike became a member of the Committee on Classification of Personnel, a group of psychologists commissioned by the U.S. Army to evaluate recruits. It was there that

Thorndike rubbed off on a young man named David Wechsler, who had just finished his master's degree in psychology. Wechsler, who would become a famous psychologist, developed a lifelong fascination with tracing the boundaries of humanity, from lower to upper limits.

In 1935, Wechsler compiled essentially all of the credible data in the world he could find on human measurements. He scoured measures of everything from vertical jump to the duration of pregnancies to the weight of the human liver and the speeds at which card punchers at a factory could punch their cards. He organized it all in the first edition of a book with the aptly momentous title *The Range of Human Capacities*.

Wechsler found that the ratio of the smallest to biggest, or best to worst, in just about any measure of humanity, from high jumping to hosiery looping, was between two to one and three to one. To Wechsler, the ratio appeared so consistent that he suggested it as a kind of universal rule of thumb.

Phillip Ackerman, a Georgia Tech psychologist and skill acquisition expert, is a sort of modern-day Wechsler, having combed the world's skill-acquisition studies in an effort to determine whether practice makes equal, and his conclusion is that it depends on the task. In simple tasks, practice brings people closer together, but in complex ones, it often pulls them apart. Ackerman has designed computer simulations used to test air traffic controllers, and he says that people converge on a similar skill level with practice on the easy tasks—like clicking buttons to get planes to take off in order—but for the more complex simulations that are used for real-life controllers, "the individual differences go up," he says, not down, with practice. In other words, there's a Matthew effect on skill acquisition.

Even among simple motor skills, where practice decreases individual differences, it never drowns them entirely. "It's true that doing more practice helps," Ackerman says, "but there's not a single study where variability between subjects disappears entirely."

"If you go to the grocery store," he continues, "you can look at the checkout clerk, who is using mostly perceptual motor skill. On average, the people who've been doing it for ten years will get through ten customers in the time the new people get across one. But the fastest person with ten years' experience will still be about three times faster than the slowest person with ten years' experience."

Scientists who study skill performance attempt to account for "variance" between people. Variance is a statistical measure of how much individuals deviate from the average. In a sample of two runners, if one athlete completes the mile in four minutes and the other runs it in five minutes, then the average is four and a half minutes and the variance is half a minute. The question for scientists is: What accounts for that variance, practice, genes, or something else?

It is a critical inquiry. It is not enough for scientists to say that practice *matters*. That point is entirely uncontroversial. As Joe Baker, a sports psychologist at York University in Toronto, says, "There isn't a single geneticist or physiologist who says hard work isn't important. Nobody thinks Olympians are just jumping off the couch."

Scientists must go beyond saying that practice matters and attempt the difficult task of determining exactly *how much* practice matters. By the strictest 10,000-hours thinking, accumulated practice should explain most or all of the variance in skill. But that never, ever happens. From swimmers and triathletes to piano players, studies report that the amount of variance accounted for by practice is generally between low and moderate.

In a study that K. Anders Ericsson himself coauthored of darts players, for example, only 28 percent of the variance in performance between players was accounted for after fifteen years of practice. At the rate of skill convergence documented in that study, a 10,000-years rule might be more likely than a 10,000-hours rule—if, that is, the players would ever reach the same level at all.

The data quite clearly support a view of skill—from chess and music to baseball and tennis—that is based on a paradigm not of "hardware *not* software," but of both innate hardware *and* learned software.

3

Major League Vision and the Greatest Child Athlete Sample Ever

The Hardware *and* Software Paradigm

In 1992, his first year of research on the Los Angeles Dodgers, Louis J. Rosenbaum met with an unexpected problem. The players were literally off the charts.

Rosenbaum had been the team ophthalmologist for the NFL's Phoenix Cardinals since 1988, and now he was in Dodgertown, the spring training facility in Vero Beach, Florida, to test eighty-seven players in the Dodgers organization, major league players as well as minor leaguers hoping to earn their spot in the show.

From eight A.M. to five P.M., Rosenbaum tested players for traditional visual acuity, dynamic visual acuity (the ability to see detail in moving objects), stereoacuity (the ability to detect fine differences in the depth of objects), and contrast sensitivity (the ability to differentiate fine gradations of light and dark). For the visual acuity test, instead of the usual eye chart with the big *E* on top, Rosenbaum and his colleagues used Landolt rings—circles with a gap in one section that the viewer must pick out as the rings get progressively smaller toward the bottom of the chart.

The trouble was that Rosenbaum used commercially available

Landolt ring charts, which tested visual acuity down to 20/15.* Nearly every player maxed out the test.

Fortunately, the other vision tests were successful. So when gruffly skeptical Tommy Lasorda, the Dodgers' legendary manager, challenged Rosenbaum to predict which minor leaguer would thrive in the majors, Rosenbaum had plenty of data to pore over. He did not have the players' baseball statistics and so had to rely purely on the vision testing data. He chose a minor league first baseman with outstanding scores.

The player was Eric Karros, a mere sixth-round pick in the 1988 draft. By '92, though, Karros was starting at first base for the Dodgers and won the National League Rookie of the Year award. It was his first of thirteen full seasons as a major leaguer.

The following spring, Rosenbaum returned to Dodgertown with a custom-made visual acuity test that went down to 20/8. Given the size and shape of particular photoreceptor cells, or cones, in the eye, 20/8 is around the theoretical limit of human visual acuity.

One's maximum visual acuity is determined by the density of cones in the macula, an oval-shaped spot in the retina of the eye. Cone density in humans is akin to the megapixel rating in digital cameras, and it is highly variable between people. Scientists who have collected retinas from deceased adults, ages twenty to forty-five, found a range from 100,000 cones/mm^2 to 324,000 cones/mm^2. (If one's cone density is below 20,000 cones/mm^2, a magnifying glass will be needed to read the newspaper.) As Michael A. Peters, author of *See to Play* and an eye doctor who works with pro baseball and hockey players, puts it: the number of cones appears to be "genetically predetermined for each of us."

Armed with a custom test at the 1993 spring training, Rosenbaum could finally measure how well pro ballplayers see. Again, Lasorda challenged Rosenbaum to predict which minor leaguer would make a

---

*Someone who scores 20/15 can stand at a distance of twenty feet and tell the difference between an *o* and a *c* that the typical person, with 20/20 vision, could only detect if they scooted up to fifteen feet.

distinguished pro. This time, the player whose vision tests stood out to Rosenbaum was Mike Piazza, a lightly regarded catcher.

Piazza had been picked by the Dodgers five years earlier in the sixty-second round of the draft, the 1,390th player taken overall, and only because Piazza's father was a childhood friend of Lasorda's. Nonetheless, Piazza would make good on Rosenbaum's prediction. He won the National League Rookie of the Year in 1993 and went on to become the greatest hitting catcher in baseball history.

Over four years of testing, and 387 minor and major league players, Rosenbaum and his team found an average visual acuity around 20/13. Position players (players who have to hit) had better vision than pitchers, and major league players had better vision than minor leaguers. Major league position players had an average right eye visual acuity of 20/11 and an average left eye visual acuity of 20/12. In the test of fine depth perception, 58 percent of the baseball players scored "superior," compared with 18 percent of a control population. In tests of contrast sensitivity, the pro players scored better than collegiate baseball players had in previous research, and collegiate players scored better than young people in the general population. In each eye test, pro baseball players were better than nonathletes, and major league players were better than minor league players. "Half the guys on the Dodgers' major league roster were 20/10 uncorrected," Rosenbaum says.

The two largest population studies of visual acuity, one from India and one from China, give a sense of just how rare 20/10 vision might be. In the Indian study, out of 9,411 tested eyes, one single eye had 20/10 vision. In the Beijing Eye Study, only 22 out of 4,438 eyes tested at 20/17 or better.

Smaller studies focused only on young people, though, have documented average vision that is better than the standard 20/20. Seventeen- and eighteen-year-olds in a Swedish study had average visual acuity around 20/16. So we should expect that Major League Baseball hitters— their average age is around twenty-eight—would have better than 20/20 vision just because they are young, but not an average of 20/11.

(Coincidentally, or perhaps not, twenty-nine often is the age at which visual acuity starts to deteriorate and the age when hitters, as a group, begin to decline.)

Mark Kipnis shared with me his first recollection of his baseball-playing son Jason's visual acuity. It was during a ski vacation when Jason was twelve years old. The Kipnis family was sitting in a large restaurant in a lodge and Mark wanted to see the score of a football game on a television in the far corner. He was tired, so he asked Jason to get up, walk over to the television, and tell him the score. "He just turned his head and told me the score," Mark says, "and a little light went off in my head." A decade later, Jason was selected by the Cleveland Indians in the second round of the 2009 draft. By 2011, he was starting at second base.

Ted Williams, the last man to hit .400 over a major league season, used to insist that he only saw ducks on the horizon before his hunting partners because he was "intent on seeing them." Perhaps. But Williams's 20/10 vision, discovered during his World War II pilot's exam, probably didn't hurt either.*

About 2 percent of the players in the Dodgers organization dipped below 20/9, flirting with the theoretical limit of the human eye. Daniel M. Laby, an ophthalmologist who worked on the Dodgers study and later with the Boston Red Sox, says that he encounters a few players at that level every year in spring training. "I can pretty comfortably say that in twenty years of caring for people's eyes I've never seen someone outside pro athletics achieve that, and I've seen over twenty thousand people," Laby says. David G. Kirschen, an optometrist who also works with professional athletes and is chief of binocular vision and orthoptic services at the Jules Stein Eye Institute at UCLA's medical school, says that he has seen a few patients outside of elite sports with 20/9 vision, "but you can count them on one hand over thirty years."

So while major league hitters might not have any faster reaction

---

*The legend that he could read the label on a spinning record, though, was a myth, according to Williams.

time than you or I do, they do have the superior vision that can help them pick up the anticipatory cues they need earlier, making raw reaction speed less important.*

Baseball players have to know before the final two hundred milliseconds of a pitch where to swing, so the earlier they pick up anticipatory cues the better. One such cue, as psychologist Mike Stadler writes in *The Psychology of Baseball*, is the "flicker" of a pitch, or the indication of the spin of the ball by the flashing pattern of rotating red seams. Two-seam fastballs and curveballs are foretold by signature red stripes on the side of the ball. A four-seam slider shows the batter a bright red dot in the center of a white circle. "That circle right out of the [pitcher's] hand, you identify in your brain, 'Oh, okay, slider,'" Keith Hernandez, the five-time All-Star first baseman, once said in television commentary of a Mets game. "If you didn't have those little red seams on the ball, you'd be in a world of trouble."

The importance of picking up ball rotation has been demonstrated in virtual-reality batting studies in which baseball players were asked to identify or to swing at digital pitches. When players picked up the rotation of the ball, they identified pitches more accurately and executed more precise swings. Hitters performed better when the red seams of the ball were accentuated, and worse when the seams were covered with white paint.

It's easy to understand why an athlete with outstanding visual acuity but without the mental database of what to look for is as useless as Albert Pujols facing Jennie Finch. But once the data is downloaded into the brain, it's advantageous to see those signals as clearly and as early as possible, all the better not to have to rely on pure reaction

---

*A study of U.S. Open tennis players also found much better visual acuity than non–tennis pros of the same age, but a few players had normal visual acuity, which suggests that excellent visual acuity is advantageous but that average vision is not an insurmountable obstacle for all tennis pros.

speed.* Al Goldis, a longtime major league scout who studied motor learning in grad school, says: "If a player has better visual skills, he can pick up the pitch while it's five feet or ten feet closer to the pitcher. If he doesn't, his mechanics might be outstanding but he reacts so late that he breaks his bat because the ball is in on his hands. It's not the bat speed, it's the visual skills. That little bit is the difference between ordinary and extraordinary."

When Laby and Kirschen studied U.S. Olympians from the 2008 Beijing Games, they found that the softball team had an average visual acuity of 20/11, outstanding depth perception, and better contrast sensitivity than athletes from any other sport. Olympic archers also had exceptional visual acuity—they scored similarly to the Dodgers—but not particularly good depth perception. That makes sense, Laby says, because the target is far away, but it's also flat. Fencers, who must make rapid use of tiny, close-range variations in distance, scored very well on depth perception. Athletes who track flying objects at a distance— softball players and to a lesser extent soccer and volleyball players— scored well on contrast sensitivity, which is "probably set at a certain ability you're born with," Laby says.†

Clearly, visual hardware interacts with the particular sports task at hand. Plus, visual hardware becomes increasingly critical the faster the ball is moving. In a study of catching skill among Belgian college students, some of whom had normal depth perception and others who had weak depth perception, there was little difference in catching ability at low ball speeds. But at high speeds, there was a tremendous difference in catching skill. Depth perception differentiated people only when the ball was whistling.

A clever follow-up study by an international team of scientists

---

*Some rare athletes simply do have superior reaction speeds. In a 1969 test, Muhammad Ali reacted to a light in 150 milliseconds, near the theoretical limit of human visual reaction time.
†There is some evidence that playing video games may improve contrast sensitivity somewhat. But they have to be action games. A relevant study found that Call of Duty 2 helped, but The Sims 2 did not.

recruited a group of young women, all with normal visual acuity but some who had poor depth perception and others with good depth perception. Each woman had a catching pretest—in which she had to snag tennis balls shot out of a machine—followed by more than 1,400 practice catches over two weeks, and then a posttest. The women with good depth perception improved rapidly during the training, while the women with poor depth perception didn't improve at all. Better hardware sped the download of sport-specific software. Conversely, a 2009 Emory medical school study suggested that children with poor depth perception start self-selecting out of Little League baseball and softball by age ten. As Gobet found with chess players, when it comes to intercepting flying objects, some catchers are more readily trainable than others.

While physical hardware alone—like depth perception or visual acuity—is as useless as a laptop with an operating system but no programs, innate traits have value in determining who will have a better computer once the sport-specific software is downloaded. Pro baseball players and Olympic softball players have outstanding vision, and Louis J. Rosenbaum was able to use tests of visual hardware to predict two straight NL Rookies of the Year—though two successes do not constitute a scientific study.

Other tests of hardware might detect the potential for greatness much earlier in life.

Psychologist Wolfgang Schneider had no idea in 1978 that he was being handed the study sample of a lifetime when the German Tennis Federation helped him and a University of Heidelberg research team recruit 106 of the most adroit eight-to-twelve-year-old tennis players in Germany.

The federation was fervent in its assistance because its officials were curious to learn whether, even among a sample restricted to kids who were already highly proficient players, the scientists could predict who might go on to be an elite adult player. Schneider's sample turned out to be quite possibly the greatest single sample of child athletes ever studied. Of 106 kids, 98 ultimately made it to the professional level, 10 rose to the top 100 players in the world, and a few climbed all the way to the top 10.

Each year for five years, the scientists gauged the children first on tennis-specific skills and then on measures of general athleticism. Schneider's expectation was that tennis-specific skills acquired through practice—like the accuracy with which a player could return a ball back to a specific target—would have predictive value for how highly ranked the children would be as adults. And he was correct. When the researchers eventually fit their data to the actual rankings of the players later on, the children's tennis-specific skill scores predicted 60 to 70 percent of the variance in their eventual adult tennis ranking. But another finding surprised Schneider.

The tests of general athleticism—for example, a thirty-meter sprint and start-and-stop agility drills—influenced which children would acquire the tennis-specific skills most rapidly. "When we omitted these motor abilities, our model no longer fit the ranking data," Schneider says. "So we said, okay, we have to keep that in our model." In other words, over the five years of the study, the kids who were better all-around athletes were better at acquiring tennis-specific skills. As with the study that examined depth perception and the ability to learn a catching skill, superior hardware was speeding the download of tennis-skill software. Schneider's study received significant attention in Germany, but because it was published in German, it garnered scant notice in the rest of the world.

Ten years later, Schneider replicated the entire study with one hundred more child tennis players. He was not nearly so fortunate with the second sample—no future world top one hundreds this time around. But the finding that general athleticism impacted tennis skill acquisition held strong. "This may not be generalizable to other sports," says Schneider, who later became president of the International Society for the Study of Behavioral Development. "But for tennis, I think it's a rather stable phenomenon."

Among the children in the original study were two, both under twelve when the testing began, who would eventually become pretty familiar in the tennis world: Boris Becker and Steffi Graf, two of the most dominant players in history. "We called Steffi Graf the perfect tennis talent," Schneider says. "She outperformed the others in tennis-specific skills and basic motor skills, and we also predicted from her lung capacity that she could have ended up as the European champion in the 1500-meters."

Graf was at the top of every single test, from measures of her competitive desire to her ability to sustain concentration to her running speed. Years later, when Graf was the best tennis player in the world, she would train for endurance alongside Germany's Olympic track runners.

The most thorough tracking of athletes from youth en route to the pros tells yet another hardware-plus-software story. As part of the "Groningen talent studies," four scientists from the University of Groningen in the Netherlands tested soccer players who were in pro-team development pipelines each year for a decade, starting in 2000 with twelve-year-old boys.

The Netherlands, despite a population of just 16.7 million, is a juggernaut in the planet's most popular team sport. The country has made the final game of the World Cup three times, including in 2010, and all of the Netherlands' professional teams have talent development programs for youth players. By 2011, sixty-eight of the hundreds of players studied had reached the professional level, nineteen of them in the Eredivisie, the premier professional league in the Netherlands.

When the study began, "I would go down on my knees and ask, 'Please can we do the testing with your players?'" recalls Marije Elferink-Gemser, of the University of Groningen's Center for Human Movement Sciences. But the work has turned out to be so valuable in predicting which players will develop best in the long term that "now clubs are coming to us and asking if we can also test their players," Elferink-Gemser says. "Now there are more clubs than we can handle."

Some of the traits that help predict the future pros are behavioral. The future pros not only tend to practice more, but they take responsibility for practicing better. Says Elferink-Gemser, "We see already when we first test them at the age of twelve that they are the players who will go up and ask the trainer, 'Why should I do this?' if they don't agree with the training."

But even among the youth soccer players—already highly pre-screened by professional clubs—small variations in physical traits at age twelve delineate the haves and the have-nots. "What we see in the shuttle sprints," Elferink-Gemser says, "is that the ones signing a professional contract later are the ones that are on average 0.2 seconds faster when they are younger, at the age of twelve, thirteen, fourteen, fifteen, and sixteen. They are always on a group average about 0.2 seconds faster than the ones who end up on the amateur level. That really gives some indication that it is important to be fast. You need a minimum speed. If you're really slow, then you cannot catch up, and speed is really hard for them to train."*

This theme isn't exactly breaking news to sports scientists. Justin Durandt, manager of the Discovery High Performance Centre at the Sports Science Institute of South Africa, is in the business of testing for speed as he scours the country for rugby players. The fastest runner he ever tested was a natural. "A sixteen-year-old boy who came from a rural area and never had a day of professional training in his life," Durandt says. The boy ran 4.68 seconds for forty meters, which would be in the 4.2-second range in the NFL-style forty-yard dash, on par with

---

*The only players who ever made up some of the sprint-speed gap were those who had not yet gone through their growth spurt—"peak height velocity," in science lingo—when they were first tested. The Groningen group tracks the height growth of players so that they can inform a coach if he's underestimating a player who simply has not yet hit puberty. Even so, markedly slow players simply never catch up, growth spurt or not.

the fastest NFL players ever. It's what Durandt hasn't seen, though, that is telling. "We've tested over ten thousand boys," he says, "and I've never seen a boy who was slow become fast."

In August 2004, a small group of scientists at the venerable Australian Institute of Sport (AIS) bet all their chips on the primacy of general, non-sport-specific athleticism.

The AIS scientists had a year and a half to try to qualify a woman for the 2006 Winter Olympics in Turin, Italy, in the winter sport of skeleton, in which the athlete begins by running down the ice with one or two hands on a sled and then, in a leap fairly like the disco move "the worm," gets on board and careens down an ice-coated track face-first on her stomach at more than seventy miles per hour.

The Aussie scientists had never even seen the sport, but they had learned that the beginning sprint accounts for about half of the variation in total race time. So they announced a nationwide call for women who could fit snugly on a tiny sled and who could sprint. Thus began Australia's Winter Olympics equivalent of *American Idol*, and it would draw commensurate media attention Down Under.

Based on written applications, twenty-six athletes were invited to the AIS in Canberra in southeastern Australia to undergo physical tests in the hope of earning one of ten funded training spots. The women came from track, gymnastics, water skiing, and surf lifesaving, a popular sport in Australia that mixes open-water rowing and kayaking, surf paddling, swimming, and footraces in the sand. Not one woman had heard of skeleton, much less tried it.

Five of the ten spots were filled solely based on the 30-meter sprint, the other five by consensus of the scientists and AIS coaches, based on how well the athletes did in a dry land test during which they had to jump on a sled fitted with wheels.

As far as the world skeleton community was concerned, the project was a doomed sideshow. "Everyone in the sport told us, 'You guys will never succeed,'" says Jason Gulbin, then a physiologist at the AIS. "They told us, 'It's a real *feel* thing. It's an art. You need time in this sport.' The biggest naysayers were really the coaches from other countries."

The women of the AIS project certainly had no feel for the ice, but they were outstanding all-around athletes. Melissa Hoar had won a world championship title in the beach-racing category of surf lifesaving. Emma Sheers had been a world water skiing champion. "It was a real curiosity," Gulbin says, "to dump basically beach babes in skeleton who had never done it before."

After selection, it was time to find out whether the women could actually get down the ice, bones intact. The scientists swallowed their nerves and headed to Calgary at the start of the winter season for the first runs on ice. It didn't take a Ph.D. to evaluate the results.

Within three slides, the newbies were recording the fastest runs in Australian history, faster than the previous national record holder, who had had years of training. "That first week on the track, it was all over," says Gulbin. "The writing was on the wall."

So much for needing a feel for the ice. Suddenly, the initial helpfulness became standoffishness as rival skeleton athletes and coaches realized they stood to be displaced or embarrassed by women they had previously viewed as rank novices.

Ten weeks after she first set foot on ice, Melissa Hoar bested about half the field at the world under-twenty-three skeleton championships. (She won the title in her next try.) And beach sprinter Michelle Steele made it all the way to the Winter Olympics in Italy.

The AIS scientists chronicled the program's success in an aptly titled paper: "Ice Novice to Winter Olympian in 14 Months."

Australia, a world sports powerhouse, has thrived off talent identification and "talent transfer," the switching of athletes between sports. In 1994, as part of the run-up to the 2000 Sydney Olympics, the country launched its National Talent Search program. Children ages fourteen to sixteen were examined in school for body size and tested for general athleticism. Australia, home to 19.1 million people at the time, won 58 medals in Sydney. That's 3.03 medals for every million citizens, nearly ten times the relative haul of the United States, which took home 0.33 medals per million Americans.

As part of the Australian talent search, some athletes were ushered away from the sports in which they had experience into unfamiliar ones that better suited them. In 1994, Alisa Camplin, who had previously competed in gymnastics, track and field, and sailing, was converted into an aerial skier. Camplin was an outstanding all-around athlete but had never even seen snow. On her first jump ever she broke a rib. On her second, she hit a tree. "Everyone thought it was a joke," Camplin told Australia's Channel Nine television network. "They told me I was too old. They told me I started too late." But by 1997, Camplin was competing on the World Cup circuit. At the 2002 Winter Olympics in Salt Lake City, despite breaking both her ankles six weeks earlier, Camplin won the gold medal. Even after that victory, watching the sparsely experienced Camplin on skis was like watching a giraffe on roller skates. She crushed her victory flowers when she fell trying to ski down the mountain to the gold medal winner's press conference.

The successes with talent transfer attest to the fact that a nation succeeds in a sport not only by having many athletes who practice prodigiously at sport-specific skills, but also by getting the best all-around athletes into the right sports in the first place. Members of the Belgian men's national field hockey team, for instance, were found to average just greater than 10,000 hours of accumulated practice, thousands more than players on the Dutch team. But the Belgian team is consistently mediocre—the Cleveland Browns of world field hockey—while the Dutch, who draw superior athletes to the sport, are a perennial world powerhouse.

The truth is, even at the most basic level, it's always a hardware *and* software story. The hardware is useless without the software, just as the reverse is true. Sport skill acquisition does not happen without both specific genes and a specific environment, and often the genes and the environment must coincide at a specific time.

Yet another remarkable finding of the chess studies of Guillermo Campitelli and Fernand Gobet was that the chance of reaching the international master level was drastically reduced if the player did not start serious chess by age twelve. It didn't necessarily matter exactly how early they started, as long as it was before twelve. Some players who start later do still reach the international master level, but their chances drop precipitously. So perhaps twelve is an approximate critical age by which certain chunks must be learned and certain neuronal connections reinforced lest the opportunity be lost.

It was once thought that as we grow and learn our brain forms neurons. But it now appears that we are born overflowing with neurons and that the ones we don't use early on are pruned away, and those that we do use are strengthened and interconnected. The brain becomes less broadly flexible but more narrowly efficient.

In his book *Why Michael Couldn't Hit*, neurologist Harold Klawans argues that, despite his transcendent athleticism, Michael Jordan was never going to learn to hit a baseball at the major league level (following his first retirement from the NBA) because the neurons he needed to learn the appropriate anticipatory skills had been pruned long ago, while he was busy playing basketball.*

This is one reason why advocates of the strict deliberate practice approach suggest that training should begin as early as possible. But it is unclear which sports truly require early childhood specialization in return for elite performance. Certainly, female gymnasts must start early. But a large and growing body of scientific evidence says that early specialization not only is *not* required to make it to the highest level in many sports, but should perhaps be actively avoided.

---

*Jordan had a .202 batting average in 127 games in AA minor league baseball. Clearly, he wasn't headed to the majors anytime soon. Still, how many adults who haven't played baseball in fifteen years could walk into AA ball, playing against former college stars and future major league pros, and hit .202? My guess is that many people would hit .000.

In sprinting, early training that is heavy and specific can be an impediment to speed development when it results in the dreaded "speed plateau." That is, the athlete gets stuck at a certain top speed and running rhythm that seems to be ingrained from early training. According to a scientific report published by the International Association of Athletics Federations (IAAF), the governing body of world track and field, "the speed plateau most often occurs in beginners who are introduced to narrowly sport-specific training too early, at the expense of general development." Says Justin Durandt, of South Africa's Sports Science Institute: "With Ericsson's 10,000-hours model, it's not that we don't believe in training, but what's happening now is that people are overtraining athletes."

A 2011 study of 243 Danish athletes found that early specialization was either entirely unnecessary or actually detrimental to ultimate development. The athletes were divided into elites, who had competed at the top level in their field, like the Olympics, and lesser, near-elites. The study focused solely on "cgs sports"—sports measured in centimeters, grams, or seconds, like cycling, track and field, sailing, swimming, skiing, and weight lifting. Both elites and near-elites "sampled" a number of sports in childhood, but near-elites—the lesser of the two groups—could be identified by a certain quality indicative of early specialization: they practiced *more* than the elites by age fifteen. It was only after age fifteen that the elites accelerated their practice pace and by age eighteen had surpassed their near-elite peers in training hours. The counterintuitive, counter-10,000-hours title of the study: "Late Specialization: The Key to Success in Centimeters, Grams, or Seconds (cgs) Sports."

The consistency of the results in those sports led South African sports physiologist and writer Ross Tucker to suggest that the elites were probably more gifted all along and simply did not have to work as hard as the near-elites early in their careers. "Their natural talent takes them to that point with less training than their peers," Tucker says. "At the age of sixteen or seventeen, when most children have matured physically, they can begin to see that they have a future in the sport and must increase training volume."*

In several popular books that give short shrift to the importance of genes, Tiger Woods is put forth as the apotheosis of the 10,000-hours model. His father facilitated colossal amounts of early childhood practice. But, by Woods's account, that was in response to his own desire to play. "To this day," Woods said in 2000, "my dad has never asked me to go play golf. I ask him. It's the child's desire to play that matters, not the parent's desire to have the child play." With Woods, one oft-omitted fact about his childhood is that, at six months old, when most infants are just beginning their struggle to stand, he could balance on his father Earl's palm as Earl walked around the house. Not to say that this necessarily destined Woods for superhuman coordination or strength as an adult, but at the very least it would seem to have given him an opportunity to start practicing earlier than other children so that he was hitting balls at eleven months. Perhaps another case of physical hardware facilitating the download of sport-specific software.

The "practice only" narrative to explain Tiger Woods has an obvious attraction: it appeals to our hope that anything is possible with the right environment, and that children are lumps of clay with infinite athletic malleability. In short, it has the strongest possible self-help angle and it preserves more free will than any alternative explanation. But narratives that shun the contributions of innate talent can have negative side effects in exercise science.

Sports scientists who do genetic work occasionally told me that their research has a public relations problem stemming from the idea that genes are rigidly deterministic, and that they negate free will or the ability to improve one's athletic station. Some genes—like the ones that give you two eyeballs or the one for the degenerative brain disease Huntington's—are rather deterministic. If you have the genetic defect

---

*A study of music students at Chetham's School of Music in England found a similar pattern. In the early stages of development, the "exceptional ability" students actually practiced consistently *less* than the "average ability" students and only later ramped up their training.

for Huntington's, you will get the disease. Many other genes, however, are not biological destiny, but simply tilt one's physical predispositions. Unfortunately, that moderate message is often entirely lost in a mainstream press that heralds each study of a new gene as if it completely supplants some aspect of human agency.

Jason Gulbin, the physiologist who worked on Australia's Olympic skeleton experiment, says that the word "genetics" has become so taboo in his talent-identification field that "we actively changed our language here around genetic work that we're doing from 'genetics' to 'molecular biology and protein synthesis.' It was, literally, 'Don't mention the g-word.' Any research proposals we put in, we don't mention the genetics if we can help it. It's: 'Oh, well, if you're doing molecular biology and protein synthesis, well, that's all right.'" Never mind that it's the same thing.

Several sports psychologists I interviewed told me that they publicly support a view that marginalizes genes because they believe it sends a positive social message. "But maybe it's dangerous too," one eminent sports psychologist told me, "to say that you're stuck where you are because you're not working hard enough." Either way, the social message has no bearing on the scientific truth.

Janet Starkes, whose work, along with Ericsson's, helped usher in the era of "software not hardware," always believed that genetic differences played a part in sport skill, but in the past she was reticent to say so publicly. "Thirty-five years ago, people very easily accepted that there are underlying innate abilities," Starkes says. "As the [learned] perceptual cognitive approach became more acceptable, it allowed me to be more centrist. It's really been very much of a pendulum swing. . . . Darts is the most closed motor skill you can get, but practice still cannot explain all the variance. And to hit [a baseball] you've got to have a modicum of visual acuity, and it's better if it's better, and you also definitely need the software to go with it."

Starkes has contributed as much to the study of skill practice as any sports scientist alive. Her work forms a full vertebra in the backbone of the strict 10,000-hours view—that only practice determines success in sports. And yet, even when she was afraid to say it, Starkes knew that without genes, the picture of sports expertise is woefully incomplete.

After all, Starkes adds, if only accumulated hours of practice matter, then why do we separate men and women in athletic competition?

It's a good question.

4

Why Men Have Nipples

Certainly, María José Martínez-Patiño never had reason to doubt her womanhood. Her face was slender and regal, its eggshell skin stretched delicately over high cheekbones. She grew up a very normal girl in northern Spain, save for being better than her peers at running and jumping.

In 1985, Martínez-Patiño, an internationally accomplished twenty-four-year-old sprint hurdler, arrived at the World University Games in Kobe, Japan, only to realize that she had forgotten the doctor's certificate that declared that she was a woman and could compete against women. So, in Kobe, she had to take the customary precompetition cheek swab to establish her biological sex.

Sex testing had been in place since the 1960s, when the International Association of Athletics Federations had seen enough brawny Eastern Bloc women—many of whom were on elaborate doping programs—that it instituted regulations to ensure that male athletes were not masquerading as females. (No such case has ever been confirmed.) Early on, testing was a crude affair. Women were made to drop their pants in front of a doctor. By the 1968 Olympics in Mexico City, that degrading procedure was replaced by tidily objective technology: swabs of cheek tissue that were tested for chromosomes. Women have XX sex chromosomes and men have XY.

Except, that is, when they don't.

Late on that August day in '85, the Spanish team doctor came to Martínez-Patiño with news. There was a problem with her test, and she would be unable to compete. Martínez-Patiño wondered whether she might have AIDS, or perhaps leukemia, which had taken the life of her brother. But the doctor would say no more.

She lived with crushing anxiety for two months. She visited doctors, but always alone, to spare her parents, who were still mourning her brother. Then the letter came. It wasn't AIDS, or leukemia, but the diagnosis would change her life. The letter said that each of the fifty cells analyzed from her cheek contained XY chromosomes. *Surprise! You're a man.* Team officials urged Martínez-Patiño to fake an injury and slink softly into retirement.

Not only did she refuse to retire, but three months later Martínez-Patiño won the Spanish national title in the 60-meter hurdles. The glory of her victory ensured her own public ridicule. The result of Martínez-Patiño's sex test was leaked to the press. The spiraling descent was swift, and cruelly thorough.

Everything that could be taken was taken. Spanish officials stripped Martínez-Patiño of her national title. They evicted her from the national athletes' living quarters. They revoked her scholarship. They expunged records of her athletic performances, as if she had never existed. Her friends sorted into those who stayed and those who fled. Her fiancé was among the latter.

Martínez-Patiño was ashamed. She lost her energy. But her resilience held fast. She maintained in the press that she was sure of her womanhood. She vowed to fight back, and help came from afar.

A Finnish geneticist named Albert de la Chapelle saw a news article about Martínez-Patiño's struggle and spoke out. De la Chapelle knew quite well that chromosomes do not necessarily make the man or woman. He had pioneered the study of individuals with XX chromosomes who develop as males. "De la Chapelle syndrome" can occur when the parents' X and Y chromosomes don't line up perfectly as they exchange information, and genes from the tip of the Y chromosome break off and end up on an X.

Martínez-Patiño paid thousands of her own dollars to be examined by doctors. They told her that she had testes, hidden from sight inside her labia, and that she had neither a uterus nor ovaries. But the doctors also discovered that, while her testes were producing male levels of testosterone, Martínez-Patiño had androgen insensitivity. That is, her body was deaf to the call of testosterone, and so she developed entirely as a woman. Most women can take advantage of the athletic benefits of the small amount of testosterone their bodies produce, but Martínez-Patiño could use none at all.

Nearly three years after her sex test became public, the Olympic Medical Commission met at the 1988 Games in Seoul, South Korea, and decreed that Martínez-Patiño should be reinstated. By that time, though, her career had been derailed, and she missed qualifying for the 1992 Olympics by one tenth of a second.

In 1990, spurred by Martínez-Patiño's ordeal, the IAAF convened an international group of scientists to decide, once and for all, how to tell a man from a woman for the purposes of competition. The experts' answer: *Don't ask us!* The group instead recommended dropping sex-verification testing altogether. By 1999, the International Olympic Committee was down to testing women only in cases where suspicion arose, and even then they had no clear standard for what constituted an eligible woman.

The trouble is that human biology simply does not break down into male and female as politely as sports governing bodies wish it would. And no technological advances of the last two decades have made the slightest difference, nor will any in the future. "I don't see how one could come up with anything different than we did twenty years ago," says Myron Genel, professor emeritus of pediatrics at Yale and a member of the group that advised the IAAF to drop the testing.

Doctors ultimately decided that Martínez-Patiño had been treated unfairly. She was, they determined, a woman for competitive purposes.

A woman with both a vagina and internal testes, breasts but no ovaries or uterus, and male doses of testosterone that circulated inertly through her body.

Neither body parts nor the chromosomes within them unequivocally differentiate male from female athletes. Is there, then, a genetic reason to separate men and women at all?

"Will Women Soon Outrun Men?" The title of the paper by a pair of UCLA physiologists seemed preposterous to me when I first saw it in 2002, as a senior in college. I had trained as an 800-meter runner for just five seasons and had already run faster than the women's world record. And I wasn't even the fastest guy on my own relay team.

But the article was in the journal *Nature*, one of the most prestigious scientific publications on the planet, so there had to be something to it. The public thought so. Of one thousand Americans surveyed by *U.S. News & World Report* prior to the 1996 Atlanta Olympics, two thirds felt that "the day is coming when top female athletes will beat top males."

The authors of the *Nature* paper graphed men's and women's world records through history for every event from the 200-meters to the marathon and saw that the improvement in women's records was far steeper than the improvement in men's. By extrapolating the curves into the future, the authors determined that women would beat men in all running events in the first half of the twenty-first century. "It is the *rates* of improvement that are so strikingly different," the authors wrote. "The gap is progressively closing."

In 2004, with the Athens Olympics as a news hook, *Nature* featured another such article, this one titled "Momentous Sprint at the 2156 Olympics?"—a reference to the projected date when women would outstrip men in the 100-meter dash.

In 2005, a paper by a trio of exercise scientists in the *British Journal of Sports Medicine* did away with the question mark and simply proclaimed in its title, "Women Will Do It in the Long Run."

Could it be that male dominance of world records was all along an artifact of discrimination that kept women from competing?

In the first half of the twentieth century, cultural norms and pseudoscience severely limited women's opportunities for athletic participation. At the 1928 Olympics in Amsterdam, the media account (which was fabricated) of exhausted female competitors lying on the ground after the 800-meter race was so distasteful to some doctors and sportswriters that the event was deemed hazardous to female health. "This distance makes too great a call on feminine strength," read a *New York Times* article.* After those Olympics, all women's events longer than 200 meters were summarily banned from the Games for the next thirty-two years. It was not until the 2008 Olympics that women finally had all the same track events as men. But as women competed in greater numbers, the *Nature* papers suggested, it looked as if they might eventually be athletically equivalent to or even better than men.

When I visited Joe Baker, a sports psychologist at York University, we discussed male/female differences in athletic performances, particularly the difference in throwing. Of all the sex differences that have ever been documented in scientific experiments, throwing is consistently one of the largest. The difference in average throwing velocity between men and women, in statistical terms, is three standard deviations. That's about twice as large as the male/female disparity in height. That means that if you pulled a thousand men off the street, 997 of them would be able to throw a ball harder than the average woman.

Baker noted, though, that the situation could reflect a lack of training in women. His wife grew up playing baseball and can easily out-throw him. "She has a laser beam," he joked. So is the difference biological?

---

*Newspapers breathlessly told of women in the 800 falling all over the track. As a 2012 article in *Running Times* reported, there was only one woman who collapsed at the finish, and three others beat the previous world record. A reporter for the *New York Evening Post* who supposedly attended the race wrote of "11 wretched women," five of whom did not finish and five of whom collapsed after the line. *Running Times* reported that there were only nine women in the race and that they all finished.

The DNA differences between men and women are extremely small, limited to the single chromosome that is either X in women or Y in men. A sister and brother draw their genes from the exact same sources—though mixing of the mother's and father's DNA, known as recombination, ensures that siblings are never close to being clones.

Much of sexual differentiation comes down to a single gene on the Y chromosome: the SRY gene, or "sex determining region Y" gene. Insofar as there is an "athleticism gene," the SRY gene is it. Human biology is set up such that the same two parents can produce both masculine sons and feminine daughters even though they're passing on the same genes. The SRY gene is a DNA skeleton key that selectively activates genes that make the man.

We all begin life as females. Every human embryo is female for the first six weeks of existence. Because mammal fetuses are exposed to a hefty dose of female hormones from the mother, it is more economical to have the default sex be female. In males, in week six, the SRY gene cues the formation of testicles and, inside them, the Leydig cells that synthesize testosterone. Within a month, testosterone is gushing and triggering specific genes to turn on and others off, and it doesn't take long for the long throwing disparity to emerge.

Boys, while still in the womb, start to develop the longer forearm that will make for a more forceful whip when throwing. And while the pronounced differences in throwing prowess are less between boys and girls than between men and women, they are already apparent in two-year-old children.

In an effort to determine how much of the throwing gap among children is cultural, a team of scientists from the University of North Texas and the University of Western Australia collaborated to test both American and Aboriginal Australian children for throwing skill. The Aboriginal Australians had not developed agriculture, instead remaining hunter-gatherers. The Aboriginal Australian girls, like the boys, were taught to throw projectiles for both combat and hunting. Indeed, the study found that throwing differences were much less pronounced between Australian Aboriginal boys and girls than between American boys and girls. But the boys still threw far harder than the girls, despite the fact that the girls were taller and heavier by virtue of their earlier maturation.

Not only are boys generally superior at throwing, but they also tend to be much more skilled at visually tracking and intercepting flying objects; 87 percent of boys outperform the average girl in tests of targeting skills. And the difference appears to be at least partly a result of exposure to testosterone in the womb. Girls who are exposed to high levels of testosterone in the womb because of a genetic condition called congenital adrenal hyperplasia, in which the fetal adrenal glands overproduce male hormones, perform like boys, not girls, on these tasks.

Highly trained women easily out-throw untrained men, but highly trained men vastly out-throw highly trained women. Male Olympic throwers heave the javelin about 30 percent farther than female Olympians, even though the women's javelin is lighter. And the Guinness World Record for the fastest baseball pitch by a woman is 65 mph, a speed routinely topped by decent high school boys. Some professional men can throw over 100 mph.

In running, from the 100-meters to the 10,000-meters, the rule of thumb places the elite performance gap at 11 percent. The top ten men in any distance—from a sprint to an ultramarathon—are about 11 percent faster than the top ten women.* At the professional level, that is a gulf. The women's 100-meters world record would have been too slow by a quarter-second to qualify for entry into the men's field at

---

*The idea that female runners surpass men as the distance of the race gets longer has been pervasive in the past. It is a topic in Christopher McDougall's fascinating *Born to Run*. But it's not quite true. The 11 percent gap among the very best performers is as firm at the longest distances as at the shortest. That said, South African physiologists found that when a man and a woman are matched for their marathon time, the man will typically beat the woman at distances shorter than the marathon, but the woman will win if the race length is extended to forty miles. They reported that this is because men are usually taller and heavier, big disadvantages the longer the race goes. Among the world's top ultramarathoners, however, the male/female size differences are smaller than in the general population, and the 11 percent performance gap exists between the best of the best of men and women in ultradistance as well.

the 2012 Olympics. In the 10,000-meters, the women's world record performance would be lapped by a man who made the minimum Olympic qualifying standard.

Larger gaps occur in throwing and pure explosion events. In the long jump, women are 19 percent behind men. The smallest gap occurs in distance swimming races. In the 800-meter freestyle, top women are within 6 percent of top men.

The papers that predicted that women will overtake men implied that the progression of women's performances from the 1950s to the 1980s was part of a stable trajectory that would continue, when in reality it was a momentary explosion followed by a plateau—a plateau that women, but not men, have reached. While women began leveling off by the 1980s in terms of top speed in events from the 100-meters to the mile, men continued to inch forward, albeit barely.

The numbers are unequivocal. Elite women are not catching elite men, nor maintaining their position. Men are ever so slowly pulling away. The biological gap is expanding.

But why does it exist in the first place?

Next to a Yellow Pages–thick dictionary on the windowsill of David C. Geary's corner office sits a woman's skull. She is overlooking the campus of the University of Missouri. "You can see the cranium is small," Geary says. He has a gaunt face and turquoise-tinted irises. A curve of gray hair that rises from the front of his forehead looks a bit like a question mark, lending his face an appropriately inquisitive air. "Her brain was only about a third the size of ours. That's why she's by the dictionary, she has to practice a lot," he jokes. Geary is referring to his scale model of the skull of Lucy, the famous *Australopithecus afarensis* ancestor of modern man whose 3.2-million-year-old bones were found in Ethiopia.

Geary spends a lot of time thinking about brains. He is a cognitive developmental psychologist, and much of his career has been devoted to understanding how children learn math, a pursuit that landed him on the president's National Mathematics Advisory Panel from 2006 to 2008. He is also a walking databank of sex differences.

Since he was a grad student at UC Riverside in the eighties, Geary has been interested in the evolution of human sex differences. But given the oft-fraught nature of research on biological sex differences— at least those that extend beyond genitalia—Geary waited until he had tenure to start publishing on human evolution. And then he exploded. He coauthored a thousand-page textbook that is nothing more than a compilation of the results of every serious scientific study of sex differences—from birth weight to social attitudes—that has been done in the last hundred years.

Though he may not have considered it before I showed up at his door, Geary's most interesting contribution to the world of sports is his 550-page tome, *Male, Female: The Evolution of Human Sex Differences*. It is the first work to incorporate all of the studies—emphasis on *all*—done on human sex differences into the framework of sexual selection.

Charles Darwin first elucidated the principles of sexual selection, though it has received far less mainstream fanfare than his other brainchild, natural selection. Whereas natural selection refers to the changes in human DNA that are preserved or eradicated in response to the natural environment, sexual selection refers to those DNA changes that spread or die out as a result of the competition for and the choosing of mates. Sexual selection is the source of most human sex differences, and it is vital to the understanding of human athleticism.

Among the physical differences between the sexes, men are generally heavier and taller and have longer arms and legs relative to their height, as well as bigger hearts and lungs. Men are twice as likely to be left-handed as women—an athletic asset in a number of sports.* Men

---

*Lefties are rare, so opponents do not face them regularly and consequently have a shallow mental database of their body movements, giving southpaws what scientists call a "negative frequency dependent advantage." In the foil fencing competition at the 1980 Moscow Olympics, for example, the entire six-man final pool was made up of lefties. French scientists Charlotte Faurie and Michel Raymond have analyzed the higher rates of left-handedness in native societies with more hand-to-hand combat. They, and others, have hypothesized that

have less fat, denser bones, more oxygen-carrying red blood cells, heavier skeletons that can support more muscle, and narrower hips, which makes running more efficient and decreases the chance of injury—like ACL tears, which are epidemic in female athletes—while running and jumping. "Because they have a broader pelvis, women have a greater angle to their knee," says Bruce Latimer, professor of anthropology and anatomy at Case Western Reserve University. "So they waste a lot of energy that goes into compression in the hip joint and it doesn't help you move forward. . . . The broader the pelvis, the more wasted energy."

One of the most pronounced physical differences between the sexes is in muscle mass. Men pack more muscle fibers into any given space in the body and have 80 percent more muscle mass in their upper body than women, and 50 percent more in their legs. As far as upper body strength, this translates to a three-standard-deviation difference in strength. That is, again, of a thousand men off the street, 997 would have a stronger upper body than the average woman.

"The differences in upper body strength are about what you see in gorillas," Geary says. "That's very big. Gorillas are the most sexually dimorphic of our close relatives. The males are about twice the size of the females. So the overall size difference is more than in humans, but the difference in upper body strength is similar."

The reason for the similarity to gorillas reflects how sexual selection has shaped human (and gorilla) athleticism. If you want to know whether the male or female of a given species is bigger and stronger, one piece of information is particularly useful: which sex has the higher potential reproductive rate.

Because of a long gestation and breastfeeding period, a female gorilla can produce only one offspring about every four years. Male gorillas collect and defend harems of females and have a much higher

---

natural selection preserves a certain amount of left-handedness, particularly in males, as a combat advantage.

potential reproductive rate. But for each male gorilla that has a harem, several other males are frozen out of breeding altogether. The result is that male gorillas compete fiercely for access to multiple females, and this "male-male competition" takes the form of fighting, or at least posturing to fight, and natural selection accentuates traits that make gorillas better fighters. "In species where females have a higher potential reproductive rate," like seahorses, Geary explains, "the situation is reversed, and the females are bigger and more aggressive." Not surprisingly, male seahorses, which care for eggs, prefer larger, stronger females.

In competition zones that are more difficult to patrol and defend physically—the sky, for example—the female's choice of a mate becomes more significant and natural selection accentuates male traits such as the attractive coloration and melodic courtship songs that occur in birds. But in primates that are confined primarily to terra firma, like gorillas and ancestral humans, head-to-head fighting can be important and evolution accentuates brute strength.

All this implies some less-than-happy notions about humans, the earth-bound primates that we are, and men in particular: that certain traits were selected for in men so that they could hurt, kill, or at least intimidate one another, and that the men who were most successful at hurting, killing, or intimidating other men sometimes used that success to mate with multiple women and to have lots of children.

The weight of evidence supports both implications. Across hunter-gatherer societies, around 30 percent of men died at the hands of other men, in combat or in raids, which often were carried out in order to capture women. As Harvard psychologist Steven Pinker put it in a talk about his book *The Better Angels of Our Nature*, about the history and modern decline of human violence: "It turns out that [Thomas] Hobbes was right. Man's life in the state of nature was nasty, brutish, and short."

The second implication, that our ancestral man strove for multiple mates, is indisputable from the genetic evidence. Because fathers pass their Y-chromosomal DNA only to sons, and only mothers pass on a type of DNA called mitochondrial DNA, we can trace our maternal and paternal ancestors separately back through time. The findings in studies throughout the world are clear: no matter where scientists look, we have fewer male than female ancestors. It took far fewer Adams than Eves to spawn the world's current population. (In some cases staggeringly so: 16 million Asian men—0.5 percent of the world's male population—have a nearly identical portion of the Y chromosome that geneticists think probably came from Genghis Khan, who famously had hundreds of wives and concubines.)

Another pattern that holds across species and among primates that have intense male-male competition is that the physical abilities important to combat are bolstered, exclusively in males, via puberty. Puberty accentuates the qualities that a burgeoning adult is soon to need for reproduction. So if athletic traits, like throwing punches or rocks, are important to reproduction, they will be magnified during puberty. And here again, men follow the violent primate pattern to a tee. Whereas girls mature early and quickly, boys go through a puberty that is both late and long, giving more time for growth, and during which their athleticism explodes.

Up until the age of ten, girls and boys have similar bodies. Girls are taller and already have slightly more body fat, but a number of athletic traits are nearly indistinguishable in boys and girls. Top running speed is almost identical in ten-year-old boys and girls, and close all the way until age fourteen, when boys very literally are on natural steroids.

At fourteen, the throwing gap, already wide, becomes a chasm. Boys develop stronger arms and wider shoulders, and by eighteen the average boy can throw three times as far as the average girl. Men also develop features that make them more difficult than boys and women to knock out: heavier brow ridges that protect the eyes and enlargement of the mandible that makes the face more resilient to blows. A glass jaw apparently did not cut it for ancestral men.

The testosterone surge of male puberty also stimulates the production of red blood cells, so men can use more oxygen than women, and it makes men less sensitive to pain than women*—just as it does to both animals and people who are given testosterone injections.

By around age fourteen, the average girl is closing in on her lifetime maximum sprint speed. World age-group records in sprinting are nearly identical for boys and girls at age nine, before puberty, when there is little biological reason for gender segregation in sports. By fourteen, however, the records are no longer in the same athletic universe.†

In some cases, women fare *worse* in certain athletic attributes after puberty. As estrogen causes fat to accumulate on widened hips, most girls experience a plateau or decline in vertical jump. And even the very leanest of adult female marathoners get down to around 6 to 8 percent body fat, double that of their male counterparts.

Studies of Olympians show that an important trait of female athletes in certain sports is that they *don't* develop the wide hips that many other women do. If elite female gymnasts go through a significant growth spurt in height or hips, their career at the top level is essentially over. As they increase in size faster than strength, the power-to-weight ratio that is so critical to aerial maneuvers goes in the wrong direction, as does their ability to rotate in the air. Female gymnasts are pronounced over the hill by twenty, whereas male gymnasts are still early in their careers. China was stripped of an Olympic gymnastics medal from the 2000 Games in Sydney when the International Olympic Committee determined that female gymnast Dong Fangxiao was two years *younger* than the minimum competition age of sixteen. It is safe to say that we will never see a similar scandal in men's gymnastics.

---

*The idea that women are more pain tolerant than men because they go through childbirth is a myth contradicted by every study done on the topic. Women are more sensitive to pain and much more likely to be chronic pain patients. Women do, however, become less sensitive to pain as they approach childbirth.

†400-meter dash records:

nine-year-old boys: 1:00.87   fourteen-year-old boys: 46.96

nine-year-old girls: 1:00.56   fourteen-year-old girls: 52.68

The advantage, then, that some female athletes have comes from certain traits that are more typical of men, like low body fat and narrow hips.

It now appears that a primary reason why women in track and field gained on men in the 1970s and '80s—and what the *Nature* papers did not account for—was because they were making up for the lack of an SRY gene by simply injecting testosterone. Beginning in the 1960s, the competition of the Cold War spilled into sports, and the systematic doping of girls, often without their knowledge, was widespread in countries like East Germany. Since that era, top women in the most explosive events have gotten worse. Seventy-five of the top eighty women's shot put throws of all time, for instance, came between the mid-1970s and 1990, predominantly from Eastern Bloc countries. That eightieth performance was a throw from East Germany's Heidi Krieger, who decades later testified in court about systematic doping in East Germany. By that time she was Andreas Krieger, having chosen to live as a man after enormous doses of steroids, which are simply testosterone analogues, pushed her body in that direction. To this day, nearly all women's world records in sprint and power events are from the 1980s, a testament to the powerful effect of male hormones on female athletes. Once the era of extreme doping ended, the performance gap between humans with and without an SRY gene stretched anew. It is now clear that the genetic advantage of men over women in most sports is so profound that the best solution is to separate them.

As Alice Dreger, professor of clinical medical humanities and bioethics in the Feinberg School of Medicine at Northwestern University and an authority on the history of sex testing in sports, told me: "The reason we have females separated in sports is because in many sports the best female athletes can't compete with the best male athletes. And everybody knows that but nobody wants to say it. Females are structured like a disabled class for all sorts of, I think, good reasons."

The difficulty in determining who is granted access to that class was evident at the 2009 world track and field championships when

Caster Semenya, a young and unheralded South African 800-meter runner, looked over her muscled shoulder and tore away from the field en route to the world title. Semenya's competitors derided her in the world media. "Just look at her," sneered Russian Mariya Savinova, the fifth-place finisher, in reference to Semenya's narrow hips and armored torso. Just looking at her, though, does not give an answer.

After the world championships, it was reported that Semenya has internal testes and no ovaries or uterus, and high levels of testosterone. (Semenya never confirmed or addressed that report.) So where, if true, should that leave her? To start breaking down sport classifications by specific biological traits, "you'd have to run international competitions like the Westminster Dog Show, with competitions for every breed," says Myron Genel, the Yale professor of pediatrics. María José Martínez-Patiño, the Spanish hurdler, had both a Y chromosome and an SRY gene, but because she was insensitive to testosterone she was ultimately allowed to compete against women.

Before the 2012 London Olympics, faced with continuing controversy over the Semenya case, the IAAF and the International Olympic Committee announced that sex would be determined based on testosterone levels. Not just the amount that is produced, but the amount the body can use.

Testosterone levels are not on a continuous spectrum. A typical woman will make less than 75 nanograms of testosterone per deciliter of blood. For men, the range is typically 240 to 1,200. So the low end of the male range is still more than 200 percent higher than the high end of the female range. In 2011, the NCAA—informed by a think tank held with the National Center for Lesbian Rights—determined that any man who undergoes sex reassignment surgery to become a woman must sit out a year while lowering her testosterone levels before she can compete on a women's team. Thus, testosterone has been deemed the source of the male athletic advantage. Though it may not be the only one.

When I spoke with endocrinologists who work with androgen-insensitive women, they all felt that XY women with androgen insensitivity—that is, like Martínez-Patiño, they can use no testosterone at all—are *overrepresented*, not underrepresented, in sports.

At the 1996 Atlanta Summer Olympics, the last that had cheek swabs, 7 women out of the 3,387 competitors—or about 1 in 480—were found to have the SRY gene and androgen insensitivity. The typical rate of androgen insensitivity is estimated to be between 1 in 20,000 and 1 in 64,000. Over five Olympic Games, an average of 1 in every 421 female competitors was determined to have a Y chromosome. So women with androgen insensitivity are vastly overrepresented on the world's largest sporting stage. Perhaps, then, something about the Y chromosome other than testosterone may be conferring an advantage.

Women with androgen insensitivity tend to have limb proportions more typical of men. Their arms and legs are longer relative to their bodies, and their average height is several inches taller than that of typical women. Like Erika Coimbra, a 5'11" Brazilian volleyball player and 2000 Olympic bronze medalist who is one of the few athletes with androgen insensitivity whose names have ever been made public. (Two of the endocrinologists I spoke with said that XY women are also overrepresented in modeling, because they are often very feminine in appearance in addition to being tall with long legs. Before her personal medical information unfortunately landed in the press, the tall, blond Coimbra had been dubbed the "Brazilian Barbie Doll.")

The increased height of XY women who are insensitive to testosterone may result from an extended growth period because they don't heed hormonal *stop* messages or from genes on the Y chromosome that influence height. Men who have an extra Y chromosome tend to be very tall. Dave Rasmussen, the tallest member of Tall Clubs International, is a 7'3" XYY male whose parents are 6'4" and 5'9".

The overrepresentation of XY women only "scratches the surface of intersex conditions in sport," as a paper in the *British Journal of Sports Medicine* put it. Jeff Brown, a Houston endocrinologist who works with some of the best athletes in America—his patients have fifteen Olympic gold medals, collectively—has treated numerous female Olympians with a condition called partial 21-hydroxylase deficiency, which can run in families and causes overproduction of testosterone.* In Brown's estimation, the condition is highly overrepresented among female athletes. "The question would be, does that put them at an advantage over someone who doesn't have it," Brown says. "Of course, the answer is yes. But that's God given. . . . I've seen it in jumpers, sprinters, and distance runners."

No scientist can claim to know the precise impact of testosterone on any individual athlete. But a 2012 study that spent three months following female athletes from a range of sports—including track and field and swimming—showed that the elite-level competitors had testosterone levels that consistently remained more than twice as high as those of the nonelites. And there are powerful anecdotes as well.†

Joanna Harper, fifty-five, is a medical physicist who was born male and later transitioned to living as a woman. Harper also happens to be a nationally accomplished age-group runner, and when she started hormone therapy in August 2004 to suppress her body's testosterone and physically transition to female, like any good scientist, she took data. Harper figured she would slow down gradually, but was surprised to find herself getting slower and weaker by the end of the first month. "I felt the same when I ran," she says. "I just couldn't go as fast." In 2012, Harper won the U.S. national cross-country title for the fifty-five-to-fifty-nine age group, but age and gender-graded performance standards indicate that Harper is precisely as competitive now as a female as she

---

*Brown has seen partial 21-hydroxylase deficiency in men as well, but the effects are less dramatic. Overall, Brown says that the endocrine systems of elite athletes differ noticeably from those of most adults. "There are all sorts of things peculiar to athletes," he says. "They are not made like me in terms of their hormonal milieu."

†Christian J. Cook, a British scientist who studies athletes and testosterone, says: "One pattern that's emerging is that top-level elite female power athletes are often closer to men in their testosterone levels . . . what those females tend to have is a great ability to garnish power from training." In a small 2013 study, Cook found that female athletes with higher testosterone self-selected more strenuous strength workouts than lower-testosterone peers.

was as a male. That is, as a female Harper is just as good relative to women as she was relative to men before her transition, but she's far slower than her own former, higher-testosterone self.

In 2003, as a man, Harper ran Portland's Helvetia Half-Marathon in 1:23:11. In 2005, as a woman, she ran the same race in 1:34:01. Harper's male time was about fifty seconds faster per mile than her female time. She has compiled data from five other runners who have transitioned from male to female, and all show the same pattern of precipitous speed decline. One runner competed in the same 5K for fifteen straight years, eight times as a man and then seven times as a woman following testosterone suppression therapy; always faster than nineteen minutes as a man, and always slower than twenty minutes as a woman.[*]

So male-typical hormone patterns (higher testosterone), skeletons (taller height, broader shoulders, denser bones, longer arms, narrower hips), and genes (SRY and others) can confer certain athletic advantages. An interesting evolutionary question, then, is: Why are women athletic at all?

Like our male forebears, our female ancestors needed to be athletic enough to walk long distances, carry kids and firewood, chop down trees, and dig up tubers. But women were far less likely to fight, to run, or to push the capacity of their upper body strength with strenuous activities like tree climbing. Part of the reason that women are as athletic as they are, Geary and several other scientists told me, might be because men are.

Consider a parallel question: Why do men have nipples? The answer is that men have nipples because women do. Nipples are abso-

*I first talked with Harper as part of the 2012 *Sports Illustrated* article "The Transgender Athlete," which I cowrote with Pablo S. Torre. Pablo and I also met Kye Allums, a former George Washington University women's basketball player and the first openly transgender NCAA Division I athlete in history. Allums had recently started testosterone injections in order to physically transition to male. He said that his hands, feet, and head had grown, his voice had become deeper, he had started growing light facial hair, and that he could run faster. Medical studies have found a dose-dependent relationship in patients between the amount of testosterone administered and the increase in muscle mass and strength.

lutely essential for reproductive success in women, and they are not so harmful in men that there has been significant natural selection pressure to get rid of them. As Harvard anthropologist Dan Lieberman, who has studied the role of endurance running in human hunting and evolution, told me, "You can't program males and females totally separately. You can't order us like a car in red or blue. Our basic biology is mostly the same, with a little difference. If women didn't need to run, you could argue that they don't need the Achilles tendon for springs in their legs. But how would you do that? You would have to have a sex-specific loss of the Achilles." Instead, nature has left humans with a system whereby—instead of great numbers of genes changing—hormones can selectively activate genes to different effect.

Men and women have almost entirely the same genes. But those small genetic differences—like the SRY gene—induce a cascade of biological consequences that lead to huge disparities on the fields of play. And not just in obvious, fixed characteristics like height and limb length. Men's muscles grow more rapidly when they lift weights than do women's. And men's hearts get bigger faster than women's in response to endurance exercise. Thus, there are small DNA differences on the Y chromosome that ultimately affect trainability.

And it's not the only chromosome with genes that do that.

5

The Talent of Trainability

His grandmother was calling him for dinner, but the boy wouldn't come. After all, he was pitching a gem, and now he was staring down the other team's slugger. This could go on for hours, the boy cleaving the air at his grandparents' house with heaters that ended in dull thuds against the rock wall.

There was no batter, of course, just a boy and his imagination, and his dream to be a pitcher. Or a catcher. Or a third baseman. Or anything, really. It needn't even be baseball. For as long as the boy could remember, he'd dreamed of being an athlete of any kind and he'd take what he could get. He just wanted to be part of a team, any team. He wasn't particularly interested in school, so how else could he distinguish himself but with his body?

One day, after watching a *Superman* episode on an old black-and-white TV, he raided the cupboards for everything from pickles to Coke and ketchup for the special shake that would give him the power to fly and thereby transform the disappointing shell he was inhabiting. But the super shake was gross, and it didn't work. Nothing did.

He couldn't make the church baseball team anymore, not since they started playing with longer baselines. He was too weak to make the throw from third base to first without bouncing it in. And despite being taller than most other kids, he was cut from the junior high basketball team. So, naturally, the boy found other ways to buoy his foundering self-esteem.

By sixth grade he was cursing and fighting. He talked back to teachers and once got kicked out of school for a day. He hid a plastic tackle box stocked with cigarettes in a bush near his house and would light up every morning before he started his paper route. He idled away hours at the bowling alley, smoking and eating junk food and learning to heist freshly baked pies from a delivery truck that always stopped out back. With that entrée into thievery, soon the boy was applying his five-finger discount to comic books and candy at the corner store. He was beginning to question the God that he had grown up with in the strict Church of Christ.

But even though the boy found his peer approval in rebellion and petty crime, there was a very conformist and very tangible something that he still yearned for: a letterman's sweater. And there was really only one sport left that he could try in junior high. Track. So he took one last crack at the Curtis Junior High School ninth-grade track team. He had auditioned in previous years, with disastrous results. He couldn't long jump, and actually managed to knock himself unconscious attempting pole vault. In seventh grade, he crashed through the hurdles, and in eighth grade he pulled a muscle in the 50-yard dash. So this time around, in the spring of 1962, he opted for the longest race the team offered, which happened to be the quarter-mile, or 400-meters; one lap around the track. Before the tryout, he asked God to please let him make this team.

When the PE teacher shouted "Go!" the boy burst to the front. He had found his calling. He was alone in the lead, his feet pounding the track like pistons with just the crackling cinders below and the blue sky in front and above. That lasted for 200 meters. And then his legs turned to bricks and sandpaper enveloped his lungs as several other boys swallowed him up and spit him out the back of their pack. He finished in just under sixty seconds, not good enough to make the team.

Still, he had led the race, however briefly. If he stuck with it, he might one day be able to run a respectable fifty-two or fifty-three seconds, he figured, and perhaps get that letter sweater. So a few times that summer before he entered high school—which started in tenth grade at Wichita East—he went outside and dashed two blocks from his house and turned around and dashed back before collapsing in the grass. And when the cross-country coach spoke at an assembly that fall, it was as if he was speaking directly at him. "Many of you boys may have done poorly in junior high sports," the coach said, "but don't be discouraged. Everyone grows at a different rate and some of you still have plenty of growing to do." The coach meant it, literally and metaphorically, so the boy went out for the team.

For his first endurance run with the cross-country team the boy was paired with Doug Boyle, another tenth grader who was blessedly likeminded and equally inexperienced. "We turned to each other and said, 'I've never run five miles before without stopping,'" the boy would recall decades later. "'Let's help each other. Let's run slow enough that we can run the whole way.'" And so they did, and they were both elated. And then the going got steeper.

In his first mile time trial, the boy ran 5:38. Not a bad start, but only fourteenth on the team. "Give it up," his concerned mother urged. "You hardly ate two bites at dinner because you didn't feel well, and you're always exhausted."

"It's too hard on you," his father said. But the boy's teammates were encouraging, and something about running had infected him. The physicality felt good to him. So he stayed out for the team, and a dramatic metamorphosis began.

In his first cross-country race, the boy was only the twenty-first best runner for the school, which landed him squarely on the C-team. But the real training had started, and he could now run ten miles without stopping. Six weeks after the start of the season, on training no different from his C-team peers, the boy moved up to junior varsity. Two months later, and to his own amazement, the boy led the varsity to the Kansas state championship.

Despite his phenomenal improvement, the boy was not set on running. "I loved the feeling of success," he would one day write, "but how I hated the pain!" So he took the winter off, thinking that perhaps a more pleasant activity might arise to occupy his spring. He daydreamed about competitive weight lifting and thought about how much he really liked golf. But come spring, he found himself out on the track. And again, while some of his peers improved by baby steps, his strides befit a giant.

That March, just six months after his 5:38 mile time trial, and despite a winter free of training, the boy ran a mile race in 4:26 to defeat the defending Kansas state champion. He followed that up with a 4:21 mile. On the team bus trip home, the coach, Bob Timmons, beckoned the boy up front to ask him just how fast he thought he could get. Perhaps 4:18 or 4:19 this year, the boy explained, and maybe 4:10 by the end of his high school career. The coach had other ideas. A decade earlier the coach had seen Roger Bannister prove to the world that a man could indeed run a mile in under four minutes without his legs crumbling to dust. And now, in the boy—in Jim Ryun—the coach saw his own little Bannister. He informed Ryun that he would become the first high school runner to dip under four. He's crazy, Ryun thought, but the seed was firmly planted.

Ryun would end tenth grade—his first track season—with a 4:08 mile. The following year, he began to train like a professional. He informed his minister that attending church three times a week was not conducive to his goal of breaking four minutes. He regularly pushed through 100-mile training weeks. He lived with his coach in the summer after one season and ran preposterously intense workouts, like forty hard intervals of a quarter-mile each. In his junior year—only his second season of track—he ran the mile in 3:59 and became a national sensation. That summer, he made the 1964 U.S. Olympic team. In 1966, as a nineteen-year-old freshman at the University of Kansas, he ran the mile in a world record time of 3:51.3. The following summer, Ryun burned up a track in Bakersfield, California, in one of the most outlandish runs in history. These days, world records in distance events almost always come in races that use "rabbits" to set the pace and cut the wind for the athlete making a record attempt. But on June 23, 1967, Ryun broke his own record without the slightest aid from a rabbit or even from his competitors, and on a cinder track, to boot. He led the race from starting gun to finishing tape in a time of 3:51.1, a mark that held for nearly eight years.

He is still remembered as one of the best middle-distance runners of all time. "Be careful those prayers that you pray!" says Ryun, who eventually traded on his athletic feats to become a Republican congressman from Kansas, thinking back on the times he asked the Lord to please at least let him make the track team. In 2007, ESPN ranked Ryun just ahead of Tiger Woods and LeBron James as the greatest American high school athlete of all time.

Without his coach planting the four-minute seed in his mind, and without his zealous training, Ryun likely would have been nothing more than an outstanding high school runner, not an athlete with his own extensive Wikipedia page. But perhaps even more than his world records, it was that 1962–63 period, before Ryun became feverishly dedicated to the goal, which stands out as especially anomalous. In that time he improved from one of the sorriest members of the high school cross country team to the best man on the state's best team, and then bettered his mile time by ninety seconds from fall to spring to the point that he was running at a pace nearly as fast as he had sprinted a quarter-mile just a year earlier. "I could not explain what was going on," he would later write of the rapid improvements. "Neither could anyone else." Not at the time, anyway.

In 1992, a collective of five universities in Canada and the United States began recruiting subjects for a seminal project known as the HERITAGE (HEalth, RIsk factors, exercise Training And GEnetics) Family Study. The universities enlisted ninety-eight two-generation families to subject their members to five months of identical stationary-bicycle training regimens—three workouts per week of increasing intensity that would be strictly controlled in the lab.

The scientists conducting the study wanted to know how regular exercise would alter these previously untrained folk. How would the strength of their hearts change? Or the amount of oxygen they could use during exercise? How would cholesterol and insulin levels fluctuate? Blood pressure would presumably go down, but how much, and would it be the same for everyone?

Unlike any previous study, DNA would be culled from all 481 participants with the goal of examining whether genes played a part in how fit one person became compared with the next. One of the prime traits of interest for the researchers was what's known as aerobic capacity, or VO_2max, in physiology lingo. Aerobic capacity is a measure of the amount of oxygen a person's body can use when he or she is running or cycling all out. It is determined by how much blood the heart pumps, how much oxygen the lungs impart to that blood, and how efficient the muscles are at snatching and using the oxygen from the blood as it hurtles past. The more oxygen one can use, the better one's endurance.[*]

Dr. Claude Bouchard, now of Louisiana State University's Pennington Biomedical Research Center and mastermind of the HERITAGE Family Study, already had an inkling of what the results would look like. In the 1980s, Bouchard had put a group of thirty very sedentary subjects through identical training plans to see how much their aerobic capacities would increase. Endurance exercise has a profound impact on the human body. More blood is produced and it flows through new capillaries that sprout like roots into muscle. The heart and lungs

---

*To be fair, VO_2max is not the sole predictor of endurance, but it is important. While knowing the VO_2max of runners in a marathon will not nearly tell you the finishing order, it may tell you which runners are professionals, which are collegians, which are weekend warriors, and which will still be running when the cleanup crews arrive. In other sports, aerobic capacity might be even more predictive. According to Swedish physiologist Björn Ekblom, data from the 1970s showed VO_2max to be a decent predictor of Olympic medals in cross-country skiing.

strengthen, and energy-generating mitochondria proliferate in the cells.

Bouchard figured he would see some variation in VO_2max improvement between people, but "the range from 0 percent to 100 percent change, I did not expect," he says. It piqued his interest enough that he decided to test identical twins in three different studies, each with a unique training protocol. Sure enough, there were high responders to training and low responders, "but within pairs of brothers, the resemblance was remarkable," Bouchard says. "The range of response to training was six to nine times larger between pairs of brothers than within pairs, and it was very consistent. That's how I was able to convince the National Institutes of Health to fund the big study, HERITAGE." It took four years to gather all the HERITAGE data, and the pattern held there too.

At each of the four centers where volunteers were made to exercise—Indiana University, University of Minnesota, Texas A&M, and Laval University in Quebec—the results of HERITAGE were astonishingly consistent. Despite the fact that every member of the study was on an identical exercise program, all four sites saw a vast and similar spectrum of aerobic capacity improvement, from about 15 percent of participants who showed little or no gain whatsoever after five months of training all the way up to the 15 percent of participants who improved dramatically, increasing the amount of oxygen their bodies could use by 50 percent or more.

Amazingly, the amount of improvement that any one person experienced had nothing to do with how good they were to start. In some cases, the poor got relatively poorer (people who started with a low aerobic capacity and improved little); in others, the oxygen rich got richer (people who started with higher aerobic capacity and improved rapidly); with all manner of variation between—exercisers with a high baseline aerobic capacity and little improvement and others with a meager starting aerobic capacity whose bodies transformed drastically.

Along the improvement curve, families tended to stick together. In other words, family members generally had similar aerobic benefits from training, while variation between different families was great. Statistical analysis showed that about half of each person's ability to improve their aerobic capacity with training was determined exclusively by their parents. The amount that any person improved in the study had nothing to do with how aerobically fit he or she was relative to others to begin with, but about half of that baseline, too, was attributable to family inheritance.

In 2011, the HERITAGE research group reported a breakthrough in exercise genetics: they identified twenty-one gene variants—slightly different versions of genes between people—that predict the inherited component of an individual subject's aerobic improvement. That still leaves the half of aerobic trainability due to other factors, but the twenty-one gene markers had the power to delineate the high and low responders. HERITAGE subjects who had at least nineteen of the "favorable" versions of the genes improved their VO_2max nearly three times as much as subjects who had fewer than ten.

Prior to this work, scientists had essentially failed to detect genes that might predict endurance improvement. A decade ago, when the sequencing of the human genome was heralded as the beginning of an age of personalized medicine, scientists hoped for a simple biological system in which a single gene or a small number of genes would define a single characteristic. Now, it is maddeningly obvious that most traits are far more complex.

The genome is a recipe book contained in every cell in the human body that tells the body how to build itself. About 23,000 pages of the book have direct instructions—or genes—for building proteins. Scientists hoped that by reading those 23,000 pages they would know everything about how the body is constructed. But the reality is that some of the 23,000 pages have instructions for an array of functions, and if one page is altered or torn out, then some of the other 22,999 pages may suddenly contain new instructions. The instructional pages, that is, interact with one another.

In the years following the sequencing of the human genome, sports scientists chose single genes that they guessed would affect athleticism and compared different versions of those genes in small groups of athletes and nonathletes. Unfortunately for that research, single genes often have tiny effects, so tiny as to remain undetectable in small studies. Even the genes for easily measured traits, like height, generally eluded detection because scientists had underestimated the complexity of genetics.

One of the innovative strokes of Bouchard and an international group of colleagues in follow-up work to HERITAGE was to let the genome tell the scientists which genes to study, as opposed to the scientists' guessing genes beforehand. In an experiment separate from HERITAGE, the group put twenty-four sedentary young men through six weeks of cycling training. They then took samples of muscle tissue from the men before and after the training program and examined which genes were more or less "expressed"—in other words, their protein-creating activity was turned up or turned down. Differences in expression levels of twenty-nine genes distinguished the high from the low responders. That is, certain genes, though present in all subjects, were more or less active in highly trainable people compared with less trainable people. The gene expression signature held true when the researchers subsequently used it to predict the training responses of a separate group of young men who were already fit and who were put on a schedule of intense interval training. (Some of the human high-responder genes also predicted exercise adaptation in rats.) Importantly, the expression levels in the twenty-nine-gene set were unaltered by exercise, indicating that those genetic expression levels constitute a genuine personal signature, not the result of prior training.

It is still unknown whether the predictor genes that Bouchard and crew have identified are the important genes themselves or whether they are simply markers for broader networks of genes. Gene expression data suggests that hundreds of genes are involved in each person's response to exercise, and that some, like the RUNX1 gene, are likely involved in changes in muscle tissue or in the formation of new blood vessels. Still others are found among genes that have helped organisms adapt to life in the oxygen-rich atmosphere of Earth that ocean bacteria started to create more than three billion years ago.

Because of the complexity of genetics, results should always be interpreted cautiously. Nonetheless, the HERITAGE findings are a stride toward understanding the genomic scaffolding of trainability, and independent work is bolstering the findings. In a separate study at the University of Miami, GEAR (Genetics Exercise and Research) researchers put 442 unrelated and ethnically diverse adults on identical cardio and weight training programs and found, just as in HERITAGE, that genes involved in the body's immune and inflammation processes predict individual differences in aerobic trainability. Some of the same genes that emerged in HERITAGE also stood out in GEAR.

When I suggested to Tuomo Rankinen, one of the HERITAGE study scientists, that some people appear to be "aerobic time bombs" awaiting training, he laughed and suggested "trainability bombs" would be a better term. It is an idea that muddles the notion of innate talent as something that appears strictly prior to training. As to the other end of the trainability spectrum, an editorial in the *Journal of Applied Physiology* noted: "Unfortunately for the low responders in these studies, the predetermined (genetic) alphabet soup just may not spell 'runner.'" There is a bright side, though, even for them.

The ultimate goal of the HERITAGE research was in line with the original promise of the Human Genome Project: to move toward personalized medicine. If doctors know how a patient responds to exercise, they can determine whether an exercise plan can usher in a desired health benefit, such as a drop in blood pressure or rise in cardiovascular power, or whether a particular low responder needs to be medicated. Fortunately, every single HERITAGE subject experienced health benefits from exercise. Even those who did not improve at all in aerobic capacity improved in some other health parameter, like blood pressure, cholesterol, or insulin sensitivity. (Then again, a

small number of exercisers with two copies of a particular gene variant actually went in the wrong direction in terms of insulin sensitivity,
suggesting that while physical activity moves most people away from
becoming diabetic, it might actually move a small number of people
closer.)

A spectrum of low to high responders appeared in every physical
quality measured, and the research team is busy looking for genes
that predict trainability for each trait. Already, genes that help account for an individual's drop in blood pressure and heart rate with
training have been identified. Variations of the CREB1 gene, which
influences the heart's pacing, were found to help predict the magnitude of the drop in a person's heart rate as he or she became more fit.

A side effect of the HERITAGE findings was to identify genetic
underpinnings, at least in that study sample, that tell the Doug Boyles
from the Jim Ryuns. Not that Boyle was a slouch. In the mile time trial
at the beginning of his senior year, he was third on the Wichita East
team in 4:39. Ryun, meanwhile, cruised to a 4:06.

By that point, though, Ryun and Boyle, initially equally intimidated
by the prospect of a five-mile run, were continents and oceans apart in
terms of their skill levels. Ryun had already been to the Tokyo Olympics
and was one of the best runners in the world. Besotted with the success
he had dreamed of back when he was making Superman shakes, and
bolstered by personal attention from his coach, Ryun made the most of
a high responder's body, pushing through 120-mile weeks and crushing workouts that are painful for most runners even to think about.
Undoubtedly, Ryun's single-minded dedication to running as fast as
possible foisted him into the athletic pantheon. But that followed on
his body's extraordinary ability to respond to training.

One can only wonder where on the HERITAGE study spectrum the
Ryun family would fall. When posed the question of whether other
Ryun family members showed signs of being very responsive to endurance training, Ryun says: "That's a good question. But I was the only
person in my family who was athletic. Nobody else was interested."

What about his younger sister? "I don't think she's run at all," he says. "She's not very gifted in that area." Then again, neither was big brother. Or so it seemed, before he started training.

This is a story that plays out on every track in America—similar boys and girls miraculously become less similar despite training similarly—albeit in far less dramatic fashion. In the absence of any biological explanation for these stories, we find other narratives to explain them, narratives that are not without consequences.

The air in the Armory Track and Field Center at 168th Street in Manhattan is notoriously stale. It was January 2002, my senior indoor track season at Columbia University, and I was not going to miss that desiccated air. At night, after races at the Armory, scratching chest pains kept me awake. Why not just skip all the training and inhale iron filings if that was going to be the payoff? But I was having a good start to the season, and on this particular day I was looking forward to testing myself against my training partner, Scott.

A moment ago, we had been warming up together, but now I'd lost him. When Scott reappeared he told me he was only going to run the first 600 meters of the race and then drop out. It was a strange choice in the final moments before the race, but one I understood.

Two years earlier, when I was a college sophomore and he was a high school senior, I hosted Scott on his recruiting trip. I knew he was a hot prospect because one of our assistant coaches gave me the "be very, very nice to this kid" talk. Despite the directive, I didn't go out of my way. Scott specialized in the same event I did, the half-mile, or 800-meters. I was a walk-on who had yet to make the varsity traveling team and was less than enthused at the thought of recruiting a phenom who was two years my junior but already running a good five seconds faster in the half-mile than my two minutes flat.

In 1997, the year I took up track as a high school junior, Scott set a fourteen-to-fifteen-year-old age group record in his home county in

Canada for the 400-meters. Not only did he appear to be talented, but he was competitive, smart, and experienced. Like other promising young runners in Canada, he had joined a club team that was more professional in its training than most U.S. high school teams. Scott just seemed like a natural. His mother had been the Canadian juvenile 100-meters champion in 1969, and she and Scott's father were the female and male track and field MVPs for the University of Windsor in 1973–74.

So why had the natural suddenly decided, before the gun had even gone off, to drop out of the race at 600 meters? Scott was struggling mentally that season. His times weren't improving, and planning to drop out was a safety valve that would release the pressure built up around him. If you drop out at 600, no one can say you failed, again, to improve your 800 time. No one can say that you have talent everyone else would kill for, but because you're not getting faster you must be a head case.

I, meanwhile, had improved relatively rapidly. I came to track late in a high school career that included football, basketball, and baseball, so I was less experienced than my recruited training partners. But, looking back, I believe that I was like a set of the HERITAGE subjects, a high responder with a low baseline.

When I first started running track in high school, I had such trouble keeping up on longer runs that I went to a pulmonologist who tested my breathing and found that I was only expelling about 60 percent as much air as my peers with each breath. Despite my youth, one of the doctor's follow-up reports notes that my result was so low as to be consistent with very early stage emphysema. When I'm in bad shape, I'm in really bad shape. As in, I get winded walking up stairs.

Each fall during college I would report to school having done the same exact, prescribed light summer training that all the half-milers did. And yet, I would invariably be in worse shape than the rest of the guys. But when the arduous training began, I would catch up, quickly. When I visited the pulmonologist in the winter, the results showed that I was miraculously transformed into a young man with the power to exhale as forcefully as my peers. Low baseline, quick responder. Every member of my training group seemed to have a higher baseline aerobic capacity, but we all responded to training to varying degrees.

In Scott's case, he would come into the season in relatively good shape and improve slowly and modestly, making it easy to brand him as a big talent who didn't capitalize on his formidable gifts. When a story like that sets in, it can be devastating, as evidenced by Scott's need to open the emergency pressure valve that day at the Armory.

I, on the other hand, was on the receiving end of a far more flattering story. I was the talentless duffer who was ready to chew through a crowbar if it meant another quarter-second off my time. Pain was nothing to me and I was making the most of my meager gifts. It was true, of course. I used to throw up after most hard practices. I would steal away to some secluded garbage bin—if I could make it in time—so that my teammates wouldn't see it.

I envied Scott when we ran side by side in practice, stealing glances at his fluid stride. But I just had to be tougher than him, I thought, because I didn't have the talent. It was an idea that coaches and teammates reinforced, as they do on every track team. I embraced the image of the hardened walk-on who squeezed drops of improvement out of a talent-dry rock of a body. When I reflect on it now, though, with the HERITAGE Family Study as my filter, I believe that the story was nothing more than a narrative that obscured a tale of genes and gene/training interactions, a tale that was playing out hidden from sight.

One day during my senior year, while searching for a cloistered nook in which to puke, I spotted Scott, already retching. And then it happened again. I saw him heaving his brains out over a trash can. And again, and again. A few times, I even saw him dart from the track halfway through a workout to throw up, and then come back and finish the intervals. Turns out, he was as tough as titanium screws. I wasn't gaining on him from the start to the finish of each season because I was outworking him. Late in my college career he and I were doing the *exact* same workouts, stride for stride. Perhaps I was gaining on him because I had a low baseline and a rapid training response. Long before I had ever even heard of HERI-TAGE or high and low "responders," I would literally start each season with the same positive self-talk: "Don't worry, they'll all be in better shape, but you respond to training like it's rocket fuel."

When one HERITAGE study scientist examined some of my genetic data, he indicated that I am likely an above-average responder to aerobic training. I also know that my blood pressure drops rapidly when I exercise. And I suspect, based on the kind of training that most benefited me in college, that I am an even greater responder to sprint-based workouts. Just as for aerobic training, low and high responders have been documented in experiments that use training programs based on explosive exercise. (If there is a lesson to be gleaned from this branch of exercise genetics, it's that there is no one-size-fits-all training plan. If you suspect that you aren't responding as well to a particular training stimulus as your training partner, you might be right. Rather than giving up, try something different.)

On that January day at the Armory, with the weight of expectation lightened by his decision not to take the race seriously, Scott ultimately decided to finish, but I blew by him with 150 meters left to run 1:54 and beat him for the first time. It was thirty seconds faster than I had run as a high school junior.

Ultimately, Scott moved away from the 800, running it less and less as his career wore on, opting for and succeeding in other events. As for me, I continued to get faster. My substantial improvement landed me a dazzling wood and glass box known as the Gustave A. Jaeger Memorial Prize, given to a four-year Columbia varsity athlete who "achieved significant athletic success in the face of unusual challenge and difficulty." Let's see someone with a high baseline aerobic capacity try to win that one.

Some people improve their endurance more rapidly than others. They are gifted with high trainability. Others are gifted with high baseline aerobic fitness. But how high can that baseline be? Or, the crucial question for sports: does anyone have elite-level aerobic endurance before training? It is a question that Norman Gledhill, a professor of kinesiology at York University in Toronto—who has administered the National Hockey League's predraft combine—began to consider in the 1970s. Gledhill's curiosity was ignited by a few cases in which a modicum of endurance seemed to precede training. The story of Nancy Tinari, a high school student at nearby George S. Henry Secondary School, was one that lodged firmly in Gledhill's mind.

In 1975, Tinari showed up to gym class in cutoff jean shorts and battered canvas Keds and proceeded, with no prior training, to cruise two miles in the twelve-minute run test. "I didn't think of myself as a jock," Tinari says. "I wasn't into equipment, or following training. I had no real interest in it." Fortunately for her, the man holding the stopwatch that fall day, George S. Gluppe, had a lot of interest in it, and "was just smart enough to realize I had someone special on my hands," he says. Gluppe pestered and prodded Tinari to start training. "You know, Nancy, you could be an Olympic runner," he told her. She laughed. Ultimately, though, Tinari gave in and made good on those words.

As soon as she started training, she started winning. After high school, Tinari ran at York and then as a professional. In 1988, despite mustering a measly thirty to thirty-five miles a week of training because of injuries, she competed in the 10K at the Olympics in Seoul. To this day, Tinari holds the Canadian national record in the 15K.

Norm Gledhill never forgot the tale of the girl who was discovered in gym class and became York University's greatest runner. He was often reminded of it through the 1980s and '90s as he and his colleague Veronica Jamnik endurance tested thousands of subjects, from elderly women to elite cyclists and rowers. Now and then, they would find someone with a VO_2max that defied their sedentary existence.

In the late nineties, Gledhill and Jamnik, along with York researcher Marco Martino, set out to see whether they could identify and study such naturally fit folk. Part of their work was to administer fitness screenings to young men hoping to become Toronto firefighters. Over two years, the team gave VO_2max tests to 1,900 young men.

Among them were six men with absolutely no history of training whatsoever who nonetheless had aerobic capacities on par with collegiate runners. The "naturally fit six," as Australian physiologists Damian Farrow and Justin Kemp call them in their sports science book *Why Dick Fosbury Flopped,* had VO_2max scores more than 50 percent higher than the average untrained young man, despite being inclined to couch-bound activities. When the York researchers examined their "hidden talents," as they call them, they saw that the naturally fit men had a crucial gift, through no discipline or effort of their own: massive helpings of blood. They were endowed with blood volumes that could have been mistaken for those of endurance-trained athletes. "It's the increased diastolic filling," explains Gledhill, referring to the part of the heartbeat when the heart muscle relaxes to allow blood back in. "When you fill up the right side of the heart with more blood, then it pumps more blood into the left side, and the left side pumps it out into the body. The [return of blood to the heart] is enhanced because of the extra blood volume."

An increase in blood volume is one of the telltale signs of a well-trained athlete. And, on occasion, a pro athlete has been caught doping with a blood volume loader in an effort to increase endurance. But not the naturally fit six; they just came out of the box that way, naturally doped.

Some of the world's greatest endurance athletes, too, seem almost to come out of the box in better shape than their peers. Athletes like Chrissie Wellington.

Wellington, a thirty-six-year-old British triathlete, made her name in Ironman races: a 2.4-mile swim, followed by a 112-mile bike ride, all before a full 26.2-mile marathon run.

She is the greatest female Ironman triathlete ever, and not by a little. In thirteen iron-distance races, including four Ironman world championships, she never lost. In July 2011, Wellington turned in one of the more outlandish performances in endurance sports history: she finished a race in Germany in 8 hours, 18 minutes, and 13 seconds; more than half an hour faster than the world record prior to Wellington's appearance in the sport in 2007. Her time would have placed her ahead of all but four men in that particular race.

By her own admission, Wellington was not interested in devoting herself to fierce sports practice when she was a child in the tiny village of Feltwell in eastern England. Her childhood passion was environmentalism. "I was the kid organizing neighborhood recycling," she says. Wellington did play sports, but "my aim in school was to get the best grades that I possibly could, and I used sports more just for fun." So she tried a bit of everything: running, field hockey, netball, and she swam for the local Thetford Dolphins.

When Wellington was fifteen, her parents realized they had an aquatic talent in their living room. Here's how Wellington recalls the conversation: "My parents said to me, 'Look, you've got potential in swimming, do you want to join the big swimming club that's an hour away and we'll drive you every morning? Or, it's your major exams coming up when you're sixteen, do you want to focus on that?' And I said, 'Look, I'd rather stay with my local swimming club, not take it so seriously, and focus on my exams.' And that's the choice I made when I was younger."

Wellington's academic focus served her well. She went on to graduate with honors from the University of Birmingham in 1998 before traveling the world and then starting on her master's degree in international development at Manchester University. In 2002, Wellington took a job at the British government's Department for Environment, Food, and Rural Affairs, or DEFRA. For two years, Wellington worked to implement development projects in impoverished countries, and she helped draft the official UK policy for the postconflict reconstruction of Iraq.

In the meantime, she had started recreational running. When she entered her first marathon, she shocked herself by finishing in three hours when her expectation had been 3:45. She was intensely passionate about her job as a civil servant, but by 2004 Wellington had tired of the bureaucratic tango it took to push incremental policy changes. She burned to have a tangible impact. So she moved to Nepal to work on a sewage sanitation project in an area ravaged by civil war. There, in the Himalayan mountains, was born the inspiration for her professional triathlon career.

Wellington had no road biking experience—she was twenty-seven the first time she sat atop a road bike—but in May 2004, just before she left for Nepal, a friend goaded Wellington to try an amateur super-sprint-length triathlon. That's a quarter-mile swim, a six-and-a-quarter-mile bike, and a one-and-a-half-mile run. Wellington borrowed a shabby road bike that she says was "black and yellow and looked like a bumblebee." Unlike the serious competitors, Wellington did not have clip-in shoes for the bike, and midway through the race she got her shoelace tangled in the gears and nearly fell. Nonetheless, she finished third and had a blast. So Wellington did two more super sprint triathlons, and won both times. When she landed in Nepal, she bought a bike.

In Nepal, Wellington would cycle with friends some mornings. Right away, she noticed, "I could just go and go and go all day." During a two-week holiday from work, Wellington and a group of friends traveled to the Tibetan capital of Lhasa and then biked eight hundred miles through the Himalayas back to Kathmandu.

Wellington had been living at an altitude of around five thousand feet in Kathmandu for eight months, so she had a degree of altitude acclimation, but much of the holiday ride was done above fifteen thousand feet—it topped out at Everest Base Camp, near eighteen thousand feet—where the air is so thin that unacclimatized folk have trouble walking, much less riding. That wasn't a problem for some of the men who rode alongside Wellington. Not only were they experienced cyclists, they were Sherpas, native Nepali people who make their living shepherding climbers up Mount Everest. "Their technical skills were far superior to mine," Wellington says, "but I could hold my own climbing up the hills and the mountains."

"When I returned to Britain from Nepal [in late 2005]," she says, "I was determined to give triathlon a good shot. I still had no idea of being professional, though."

In February 2006, shortly after her return, Wellington was at a wedding in New Zealand and "got roped" by friends into entering a 151-mile running, cycling, and kayaking adventure race through the Southern Alps. Wellington's lifetime of kayak training consisted of a crash-course tutorial the previous month. Despite capsizing several times in the kayak leg of the race, she placed second. In September, juggling training and a full-time job, Wellington won an amateur triathlon world championship title. Five months later, in February 2007, she turned pro.

In October that year, despite having trained exclusively for shorter triathlon races, Wellington entered the Ironman World Championship as a virtual unknown to her competitors. Her anonymity lasted through the early afternoon of October 13, 2007, when she arrived at the running portion of the world championship race two minutes ahead of the closest woman. "I kept expecting the other athletes to be stronger and come up and pass me," Wellington says, "but the gap was just growing." By the finish line, the gap had grown to five minutes.

The British Triathlon Federation hailed the win as "a remarkable feat, deemed to be a near impossible task for any athlete racing as a rookie at their first Ironman World Championships." Wellington bested athletes like second-place finisher Samantha McGlone. In each of the five previous years, while Wellington was helping bring drinking water to third-world countries, McGlone had been a member of the professional Canadian National Elite Triathlon Team. She had already competed in triathlon at the 2004 Athens Olympics, and, unlike Wellington, was actually deliberately practicing to race the Ironman distance. "We all have talents," Wellington says. "And sometimes those talents are hidden and you have to dare to try something new or you might not know what you're good at."

By the time Wellington retired in December 2012—ending a career that was five years from start to finish—triathlon-as-after-work-hobby was a receding memory. As a professional, Wellington trained zealously. Six sessions each of swimming, biking, and running per week and six-hour training days were not uncommon, not to mention the massage afterward and meticulously planned meals and sleep. She never stopped improving throughout her career and her best work may still have been ahead. But it was her rapid rise that was most startling.

When asked what her weak spot was relative to her competitors, Wellington is quick to point to her swimming, the discipline in which, interestingly, she has the deepest experience.

In the York University study, 6 out of 1,900 men were in the naturally fit fraternity. That sounds rare at first blush, but 6 out of 1,900 suggests that most large high schools have a few naturally fit boys, and, if the results can be extrapolated to women, there are more than 100,000 naturally fit people in the United States between the ages of twenty and sixty-five. Looking at it this way, it's reasonable to wonder whether every professional endurance athlete in history might not have started as a member of the naturally fit fraternity.

The school fitness test—like the one that set Nancy Tinari on an Olympic path—is not a terribly uncommon venue for the identification of future world-beaters. Meb Keflezighi, the Eritrean-American who in 2009 became the first American man to win the New York City Marathon in twenty-seven years, initially got a sense of his endurance prowess during the mile run in seventh-grade gym class in San Diego. "I just ran hard because I wanted that A; I had no idea of strategy or pace," Keflezighi writes in his autobiography, *Run to Overcome*, of the 5:10 mile he clocked on no training. The gym teacher phoned the San

Diego High School cross-country coach and told him, "We've got an Olympian here." Indeed. Keflezighi went on to win the silver medal in the marathon at the 2004 Olympics in Athens. "A PE class had turned my life around," he writes, "though I didn't know it at the time."

Andrew Wheating, a twenty-five-year-old American and one of the nation's top milers, did not run his first track race until his senior year at tiny Kimball Union Academy in Meriden, New Hampshire. His running career was kick-started when he ran a five-minute mile as part of a conditioning drill before his junior-year soccer season. Apparently sensing that his athletic future lay on the track and not the pitch, Wheating's soccer coach suggested he switch to cross-country. Wheating did so and earned a track scholarship to the University of Oregon, a powerhouse running program. The summer after his sophomore year at Oregon—his third track season ever—Wheating made the U.S. Olympic team in the 800-meters. Two years later, at the end of the 2010 track season, Wheating ranked fourth in the world in the 1,500-meters with a time of 3:30:90, equivalent to a mile time under three minutes and fifty seconds.

Cuban runner Alberto Juantorena, who in 1976 became the only athlete ever to win gold medals in both the 400- and 800-meters, had been an aspiring basketball player in 1971 when the national team basketball coach suggested he switch to running. "Thanks for the offer, but no, I'd rather not," Juantorena insisted. "You know basketball is my life." To which the coach responded: "We're sorry, but it's already been decided that you will change sports. Starting tomorrow, you are a runner, not a basketball player." Juantorena made the Munich Olympics the very next year.

But it could also be that some naturally fit people are not like Wellington or Wheating, but are instead like the low responders in the HERITAGE study, in which case they would not improve rapidly with training. (Bouchard's research group also has DNA from three hundred endurance athletes with very high VO_2max scores. Based on their gene variants, unsurprisingly, none of them is predicted to be at the low end of the responder spectrum.) Based on his data, Bouchard estimates that between one in ten and one in twenty people start with elevated aerobic capacity—though not nearly as high as the naturally fit six—and between one in ten and one in fifty people are high aerobic responders. "The probability that a person will be highly endowed and highly trainable is the product of those two probabilities," Bouchard says. "It's not pretty. It's between one in one hundred and one in one thousand."

The ultimate combination, of course, would be a person who starts with a highly elevated aerobic capacity and has a rapid training response. It is very difficult to identify those people before they start training, as athletes are not normally subjected to lab tests until they have already accomplished something. Science is far better at looking at an elite athlete and retrospectively suggesting why that individual is succeeding than in finding someone who might succeed before he or she has started practicing—and had a chance to respond to training—and then following them.

But there was a bit of unique and relevant science done by Dr. Jack Daniels, an exercise physiologist, former U.S. Olympic pentathlete, and one of the world's most well-respected endurance coaches. Decades ago, Daniels tracked one Olympic runner over the course of five years, testing him on all manner of biological traits at least every six months. When fully trained, the runner had a VO_2max about double that of an average, untrained but healthy man. The third year of the study, however, brought an unexpected research problem: the athlete got sick of competing. The pressure of expectations and the ceaseless slog and drudgery of interval training got to him. The runner was less than halfway through a race at the national championship when he simply stepped off the track and refused to run another step for an entire year. More than a year and a half passed before he really got serious about racing again.

Instead of giving up on his research, Daniels tested the athlete during his year of lazing around. An athlete who stops training can within weeks lose more than 15 percent of the VO$_2$max he built up. The runner's VO$_2$max had fallen 20 percent by the time Daniels tested him. In the absence of training, the Olympian's aerobic capacity lined up exactly with the naturally fit six from the York University study. (Decades later, this would become a familiar pattern to Daniels. In 1968, for his Ph.D. dissertation, Daniels tested twenty-six elite runners, fifteen of whom went on to the Olympics. When he retested them in 1993, even those who had stopped running many years earlier and were overweight maintained a VO$_2$max much higher than normal men. Said Daniels, in an interview with *Flotrack*: "Even the ones who hadn't [continued to] run pretty well proved the genetic characteristics.")

After a year of mental convalescence, the runner started jogging with his wife. And with the Olympics approaching, his fire for full-time training was reignited. As the runner gradually increased his training intensity, he quickly regained, essentially precisely, the 20 percent of his aerobic capacity that he had lost during his fallow period.

From a physiological standpoint, what Daniels chronicled over five years was in line with the results of a seven-year study of male, junior Japanese distance runners. The boys were selected for the study because each had won a middle- or long-distance event at the Japan Junior Championship. The study then followed the boys as they trained assiduously from the ages of fourteen to twenty-one for two hours each day, five or six days a week. Their aerobic capacities started at almost the identical level as Daniels's Olympian during his period without training—around the same level as the naturally fit six. Over the course of their years of training, the boys all improved, but naturally divided into two groups: study group I, which saw an average aerobic capacity increase of 13 percent; and study group II, with boys who hit a plateau of 9 percent aerobic improvement—as well as a plateau in race time improvement—by age seventeen. Each of the boys in the latter group, having ceased to improve, quit running altogether after age seventeen. There may have been, in effect, a form of natural self-selection that left in the competitive population (and working toward their 10,000 hours) only those boys who continued to improve. That is not to say the boys who continued competing were just lucky. The study suggests that the more potential to improve the boys had, the longer and harder they had to work to reach it. But the ability to improve may have kept them in the sport and dedicated to training.

So the group I Japanese boys appeared, like Daniels's Olympian, to have a high baseline aerobic capacity as well as the ability to improve more than their peers. But Daniels's Olympian improved even more, while some of his peers, like the group II Japanese boys, flat-lined and went off to follow other interests. From the looks of it, Daniels's Olympian both had a naturally high aerobic fitness and was a high responder to training.

Incidentally, the Olympian was Jim Ryun.

6

Superbaby, Bully Whippets, and the Trainability of Muscle

The baby boy was born around the turn of the millennium, and it was the twitching that grabbed the nurse's eye. Sure, the boy was slightly on the heavy side, but nothing jaw-dropping for the nursery at Charité hospital in Berlin. But those jitters. The little ticks and shudders that started just a couple of hours after he was born. The doctors worried that he might have epilepsy, so they sent him to the neonatal ward. That's where Markus Schuelke, a pediatric neurologist, noticed his pipes.

The newborn had slightly bulging biceps, as if he had been hitting the womb weight room. His calves were chiseled, and the skin over his quads was stretched a bit too taut. Soft as a baby's bottom? Not this baby. You could bounce a nickel off these glutes. Ultrasound examination of his lower body showed that the boy was beyond the top of the baby charts in the amount of muscle he had, and beneath the low end of the charts in terms of fat.

The boy was otherwise normal. The functioning of his heart was ordinary, and the jitters subsided after two months. *Perhaps the baby was the Benjamin Button of bodybuilding, and would gradually lose muscle.* Not quite. By the age of four, he had no trouble holding 6.6-pound dumbbells suspended horizontally at arm's length. (Imagine toddler-proofing that household.)

Monstrous strength ran in the family. The boy's mother was strong, as were her brother and father. But her grandfather—he was acclaimed on his construction crew for unloading 330-pound curbstones from truck beds with his bare hands.

Fully clothed, the boy did not stand out from his peers. You wouldn't ogle his puerile pecs if you passed him in the street. But the muscles in his upper arms and legs were roughly twice the size of other boys his age. *Double muscle.* It reminded Schuelke of something.

In the early 1990s, Johns Hopkins geneticist Se-Jin Lee had begun searching for muscle in his lab on North Wolfe Street in Baltimore. Not the finished muscle tissue itself, but the protein scaffolding that builds it. The purpose of the search was to find treatments for muscle-wasting diseases, like muscular dystrophy. Lee and a group of colleagues targeted a family of proteins known as transforming growth factor-ß. They cloned genes that coded for the proteins and then set off like kids with new toys, trying to figure out what the heck each gene did.

They gave the genes prosaic names—growth differentiation factor 1 through 15—and then bred mice that lacked working copies of each gene, one at a time, so they could see what would happen and thereby deduce each gene's function. The mice without GDF-1 had their organs on the wrong side. They didn't survive long. The mice without GDF-11 had thirty-six ribs. They, too, died quickly. But the mice without GDF-8 survived. They were freak show rodents of a different kind. They had *double muscle.*

In 1997, Lee's group named GDF-8, a gene on chromosome two, and its protein "myostatin." The Latin *myo-*, meaning muscle, and *-statin*, to halt. Something that myostatin does signals muscles to cease growing. They had discovered the genetic version of a muscle stop sign. In the absence of myostatin, muscle growth explodes. At least it did in the lab mice.

Lee wondered whether the gene might have the same effect in other species. He contacted Dee Garrels, owner of the Lakeview Belgian Blue Ranch in Stockton, Missouri. Belgian Blue cattle are the result of post–World War II breeding that sought more meat to accommodate the increased demand of Europe's postbellum economy. Breeders in Belgium crossed Friesian dairy cows with stocky Durham shorthorns and got cattle with heaps of muscle. Double muscle, to be precise. Belgian Blues look as if somebody unzipped them and tucked bowling balls inside their skin. "Hotline," Garrels's 2,500-pound prize Belgian Blue bull, once ripped a steel restraining gate off its hinges and flicked it aside en route to a cow in heat.

Lee asked Garrels for blood samples from her double-muscled cattle. Sure enough, the Belgian Blues were missing eleven of the DNA base pairs—out of more than six thousand—from the myostatin gene. It left them without a stop sign for their muscles. Another breed of double-muscled cattle, Piedmontese, also had a genetic mutation that resulted in no functional myostatin.

So Lee went hunting for human subjects. First stop: the grocery store, where he loaded his cart with muscle mags, the kind with cover photos of bulging-veined men in itty-bitty skivvies. Lee has jokingly been called "the skinniest man in the world" by a colleague, and he still remembers the sideways look from the cashier. Nonetheless, he placed an ad in *Muscle and Fitness* and was immediately swamped with willing volunteers, many of whom mailed him photos of themselves flexing and scantily clad, or not clad at all. He took samples from 150 muscular men, but found no myostatin mutants.

He put the work aside until 2003, when Markus Schuelke called to talk about the bulging baby boy who was born at Charité hospital three years earlier and whose development he was monitoring. The following year, Schuelke, Lee, and a group of scientists published a paper that would introduce the world to the "Superbaby," as the media would name him. The German boy, whose identity has been carefully guarded, was the human version of a Belgian Blue. Mutations on both of his myostatin genes left him with no detectable myostatin in his blood. Even more provocatively, Superbaby's mother had one typical myostatin gene and one mutant myostatin gene, leaving her with more myostatin than her son but less than the average person. She was the only adult with a documented myostatin mutation, and she was a professional sprinter.

Double muscle might seem like an unconditional blessing, but myostatin exists for a reason. It is, in evolutionary terms, "highly conserved." The gene serves the same function in mice, rats, pigs, fish, turkeys, chickens, cows, sheep, and people. This is probably because muscle is costly. Muscle requires calories and specifically protein to sustain it, and having massive muscles can be a massive problem for organisms—like ancestral humans—that don't have steady access to the protein necessary to feed the organs. But that is a diminishing concern in modern society.

In Superbaby's case, doctors initially worried that the boy's lack of myostatin might cause his heart to grow out of control. So far, though, no major health concerns have been reported in him or his mother.* Thus, it seems unlikely that an individual with a myostatin mutation would ever even think to get tested. The result is that nobody has any idea just how rare the myostatin mutation is, other than that most people (and animals) don't have it. But the facts that the one boy with two of the rare myostatin gene variants has exceptional strength, and that his mother had exceptional speed, are no coincidence. Superbaby and his mother fall precisely in line with racing whippets.

Since the late nineteenth century, speed-seeking whippet breeders unknowingly created dogs that, like Superbaby's mother, have single myostatin mutations and that are lightning fast. In the highest level of

---

*A decline in myostatin is actually a normal adaptation among people who lift weights regularly, apparently part of the body's way of clearing the road for muscle building.

whippet competition—racing grade A—where dogs hit a top speed of thirty-five miles per hour, more than 40 percent of the dogs have what is normally an exceedingly rare myostatin mutation. In racing grade B only about 14 percent have it. In grade C, it all but disappears.

Even in racing grade A, the myostatin mutation is not a prerequisite, but it is clearly beneficial. The drawback in the whippet breeding system is that some dogs end up with too much muscle.

Every whippet puppy inherits one copy of its myostatin gene from each parent. If two sprinter whippets—dogs that each have one copy of the myostatin mutation—have four puppies, this is the likely scenario: one puppy will have zero copies of the mutation and be normal; two puppies will have one copy of the mutation, like Superbaby's mother, and be sprinters; the fourth puppy will have two copies of the mutation, like Superbaby, which make for a double-muscled "bully" whippet. Bully whippets are cartoonishly buff. A bully whippet looks like a shrink-wrapped rock pile stuck to a cuddly face. Bully whippets are too bulky to sprint, so breeders often put them down.

The more scientists look, the more species they find that hold with the pattern of myostatin gene mutations and speed. In early 2010, two separate studies independently found that variations in the myostatin genes of Thoroughbred racehorses were powerful predictors of whether the horses were sprinters or distance runners. And horses with a so-called C version of the myostatin gene—a variant that results in less myostatin and more muscle—earned five and a half times more money in purses than did their counterparts that carried two T versions and were flush with myostatin.

The scientists who discovered this have, not surprisingly, started genetic testing companies for Thoroughbred breeders.

As soon as Lee published his first mighty mice results in 1997, he was inundated with messages from parents of children with muscular dystrophy (no surprise), and also from athletes (surprise!) ready and willing to offer themselves for genetic experimentation. Some of the athletes hardly knew what they were talking about. They asked Lee where they could purchase myostatin, not realizing that it is the absence of myostatin that leads to muscle growth.

Lee himself is a huge sports fan. He can recite the last forty-five NCAA basketball champions and he reverse-engineers the memory of a two-decades-old date with his wife by first thinking of who the St. Louis Cardinals' pitcher was that day. But he has been reticent to talk with sportswriters about his work. He is troubled by the apparent willingness of athletes to abuse technology that isn't even technology yet, and that is meant for patients with no other options. He hopes that any future myostatin-based treatments won't be stigmatized the way steroids have been because of their role in sports scandals.

The occasional keyhole view into the genetic cutting edge has proven understandably tantalizing for athletes. After myostatin, Lee moved on to mice in which he both blocked myostatin and altered another protein involved in muscle growth, follistatin. The result: *quadruple muscle.* In collaboration with researchers at the pharmaceutical company Wyeth, Lee then developed a molecule that was shown to bind to and inhibit myostatin and with just two injections increased mouse muscle 60 percent in two weeks. A subsequent trial by the pharmaceutical company Acceleron reported in 2012 that a single dose of that same molecule boosted muscle mass in postmenopausal women. Several companies now have myostatin inhibitor drugs in clinical trials.

For pharmaceutical companies, this is the search not simply for a cure for muscle-wasting diseases, but for the all-time pharma pot of gold: a cure for the normal muscular decline of aging. And myostatin is not the only gene that has emerged in the quest for explosive muscle growth.

The year after Lee's mighty mice made headlines, H. Lee Sweeney, a physiology professor at the University of Pennsylvania, introduced the world to his own ripped rodents, made by injecting them with a transgene—a gene engineered in a lab—to produce the muscle-building insulin-like growth factor, or IGF-1. Like Lee, Sweeney was flooded with calls. A high school wrestling coach and a high school football coach both offered up their teams as genetic guinea pigs. (Of course, the offers were rejected.)

The gene-doping era may even already be here. In 2006, during the trial of German track coach Thomas Springstein on charges of providing performance-enhancing drugs to minors, evidence emerged that the coach had been seeking Repoxygen, an anemia drug that delivers a transgene that prompts the body to produce red blood cells.

Before I traveled to the Beijing Olympics in 2008, a former world powerlifting champion gave me the name of a Chinese company that he said bodybuilders were using for gene therapy techniques. Once in China, I contacted the company, and a representative did respond to discuss potential genetic technologies. But I suspect it was just a strategy to tantalize patients, and that the company was not actually performing gene therapy.

Still, Sweeney says that one method of delivering transgenes, simply pouring them into the bloodstream, is not necessarily safe but is simple enough that it could be accomplished by a sharp undergrad studying molecular biology. Sweeney has helped World Anti-Doping Agency officials prepare to fight gene doping, but if gene therapy is proven completely safe, he says, his impetus for keeping it out of sports disappears.*

But perhaps the most interesting question is whether common DNA sequence variations in genes like IGF-1 and myostatin, as opposed to rare mutations, help to determine whether one gym-goer will pack on muscle faster than her spotting buddy. Comparisons of common variants of the human myostatin gene in both weight lifting

*In a famous gene-therapy trial in France, twelve boys were successfully treated for X-linked severe combined immunodeficiency—colloquially known as Bubble Boy syndrome—but several of the boys subsequently developed leukemia.

and sedentary subjects have had less than eye-popping results. Some studies have found slight differences and some none at all. Other genes involved in the process of muscle building, though, are emerging as critically important to understanding why some people get sculpted when they pump iron while others struggle toward buffness in vain.

Muscles are pieces of meat made of millions of tightly packed threads, or fibers, each a few millimeters long and so thin as to be barely visible on the end of a needle. Along each fiber are a number of command centers, or myonuclei, that control muscle function in the area. Each command center presides over its fiber fiefdom.

Outside of the fibers hover satellite cells. These are stem cells that wait quietly, until muscle is damaged—as happens when one lifts weights—and then they swoop in to patch and build the muscle, bigger and better.

For the most part, as we gain strength we do not gain new muscle fibers but simply enlarge the ones we already have. As a fiber grows, each myonuclei command center governs a larger area, until the point when the fiber gets big enough that the command center needs backup. Satellite cells then form new command centers so the muscle can continue to grow. A series of studies in 2007 and 2008 at the University of Alabama–Birmingham's Core Muscle Research Laboratory and the Veterans Affairs Medical Center in Birmingham showed that individual differences in gene and satellite cell activity are critical to differentiating how people respond to weight training.

Sixty-six people of varying ages were put on a four-month strength training plan—squats, leg press, and leg lifts—all matched for effort level as a percentage of the maximum they could lift. (A typical set was eleven reps at 75 percent of the maximum that could be lifted for a single rep.) At the end of the training, the subjects fell rather neatly into three groups: those whose thigh muscle fibers grew 50 percent in size; those whose fibers grew 25 percent; and those who had no increase in muscle size at all.

A range from 0 percent to 50 percent improvement, despite identical training. Sound familiar? Just like the HERITAGE Family Study, differences in trainability were immense, only this was strength as opposed to endurance training. Seventeen weight lifters were "extreme responders," who added muscle furiously; thirty-two were moderate responders, who had decent gains; and seventeen were nonresponders, whose muscle fibers did not grow.*

Even before the strength workouts began, the subjects who would ultimately make up the extreme muscle growth group had the most satellite cells in their quadriceps, waiting to be activated and build the muscle. Their default body settings were better primed to profit from weight lifting. (Incidentally, one possible reason steroids help athletes gain muscle rapidly is because the drugs prompt the body to make more satellite cells available for muscle growth.)

Every similar strength-training study has reported a broad spectrum of responsiveness to iron pumping. In Miami's GEAR study, the strength gains of 442 subjects in leg press and chest press ranged from under 50 percent to over 200 percent. A twelve-week study of 585 men and women, run by an international consortium of hospitals and universities, found that upper-arm strength gains ranged from zero to over 250 percent.

The results evoke the American College of Sports Medicine's new motto: "Exercise Is Medicine." Just as areas of the genome have been identified that influence how well different people respond to coffee, Tylenol, or cholesterol drugs, so does every individual seem to have a physiologically personalized response to the medicine of any particular variety of training.

The Birmingham researchers took a HERITAGE-like approach in

---

*It's important to keep in mind that the harder the training, the less likely there are to be "nonresponders." The harder the work, the more likely a subject will get at least some response, even if it is less than her peers.

their search for genes that might predict the high satellite cell folk, or high responders, from the low responders to a program of strength training. Just as the HERITAGE and GEAR studies found for endurance, the extreme responders to strength training stood out by the expression levels of certain genes.

Muscle biopsies were taken from all subjects before the training started, after the first session, and after the last session. Certain genes were turned up or down similarly in all of the subjects who lifted weights, but others were turned up only in the responders. One of the genes that displayed much more activity in the extreme responders when they trained was IGF-IEa, which is related to the gene that H. Lee Sweeney used to make his Schwarzenegger mice. The other standouts were the MGF and myogenin genes, both involved in muscle function and growth.

The activity levels of the MGF and myogenin genes were turned up in the high responders by 126 percent and 65 percent, respectively; in the moderate responders by 73 percent and 41 percent; and not at all in the people who had no muscle growth.

The network of genes that regulates muscle growth is only beginning to be delineated, but one biological cause of individual differences in strength building is already well-known. Some athletes have greater muscle growth potential than others because they start with a different allotment of muscle fibers.

Coarsely speaking, muscle fibers come in two major types: slow-twitch (type I) and fast-twitch (type II). Fast-twitch fibers contract at least twice as quickly as slow-twitch fibers for explosive movements—the contraction speed of muscles has been shown to be a limiting factor of sprinting speed in humans—but they tire out very quickly.*

---

*Slow-twitch muscle fibers require abundant oxygen, and thus are surrounded by blood vessels, which makes them appear dark. At Thanksgiving dinner, you can tell that turkeys are predominantly walkers, not fliers, because the dark meat is in the legs, and white, fast-twitch

Fast-twitch fibers also grow twice as much as slow-twitch fibers when exposed to weight training. So the more fast-twitch fibers in a muscle, the greater its growth potential.

Most people have muscles comprising slightly more than half slow-twitch fibers. But the fiber type mixes of athletes fit their sport. The calf muscles of sprinters are 75 percent or more fast-twitch fibers. Athletes who race the half-mile, as I did, tend to have a mix in their calves closer to 50 percent slow-twitch and 50 percent fast-twitch, with higher fast-twitch proportions at the higher levels of competition. Long-distance runners are skewed toward the slow-twitch muscle fibers that can't produce explosive force as quickly, but which tire very slowly. Frank Shorter, the last American man to win the Olympic marathon, was found to have 80 percent slow-twitch muscle fibers in a leg muscle that was sampled. It begs the question of whether the athletes get their unique muscle fiber combinations via training or whether they gravitate to and succeed in their sports because of how they're already built.

A vast body of evidence suggests that it is more of the latter. No training study ever conducted has been able to produce a substantial switch of slow-twitch to fast-twitch fibers in humans, nor has eight hours a day of electrical stimulus to the muscle. (That caused a fiber type switch in mice, but failed to do so in people.) A 2010 review of muscle fiber type studies in the *Scandinavian Journal of Medicine & Science in Sports* had this answer in response to the question of whether significant fiber type switches can occur through training: "The short (disappointing) answer is, 'Not really.' The long answer has some uplifting nuances."* Meaning that aerobic training can make fast-twitch fibers more endurant and strength training can make slow-twitch fibers stron-

---

meat is in the breast. The slow-twitch fibers are iron-rich, so if you're looking to add iron to your diet, go for the turkey legs.

*A 2009 study of 1,423 Russian endurance athletes and 1,132 nonathletes found a moderately strong and highly statistically significant correlation between the proportion of slow-twitch fibers a subject had and the subject's versions of ten different genes that independent studies had associated (albeit often tenuously) with endurance. Little is known, though, about the specific genes that influence fiber-type proportions.

ger, but they don't completely flip. (Save for extreme circumstances, like if one's spinal cord is severed, in which case all fibers revert to fast-twitch.)

Both gene and fiber type data suggest that innate qualities of each individual ensure that there is no one-size-fits-all sport or method of training. Some sports scientists have already put that notion to practical use.

With just 5.5 million residents, Denmark cannot afford to squander its top athletes. So Jesper Andersen makes sure that Danish athletes and coaches are thinking about muscle fiber type.

Andersen was a national-level 400-meter runner and later coached the Danish national team sprinters. Now he is a physiologist at the world-renowned Institute of Sports Medicine Copenhagen. He works with elite athletes ranging from Olympic runners to soccer players on Denmark's best team, F.C. Copenhagen, which competes in Europe's Champions League. And he sees individualized responses to training programs every day.

When Andersen took muscle biopsies of Danish shot-putters in 2003, he found that Joachim Olsen had a much higher proportion of fast-twitch fibers in his shoulders, quads, and triceps than the other top throwers. Andersen became convinced that Olsen had not nearly reached his muscle growth potential, given his high proportion of fast-twitch fibers. So he urged Olsen to stop weight training throughout the year and instead to focus on shorter periods of extremely heavy weight lifting, followed by periods of total rest with no weight lifting at all. Over the course of one season, Olsen's muscle fibers ballooned—as confirmed by another biopsy—and the following summer he won the bronze medal at the 2004 Olympics in Athens. The feat propelled him to celebrity status in Denmark, and he subsequently won the Danish version of *Dancing with the Stars* and was elected to parliament.

In the shoulder muscle of one Danish national team kayaker (as well as in that kayaker's brother) Andersen found more than 90 percent slow-twitch muscle fibers. The kayaker was trying to qualify for the Olympics in either the 500- or 1,000-meter race, but his competitors were so much more explosive off the starting line that even though he always caught up late in races, he constantly fell short and had no chance to make the Olympic team. Andersen told the kayaker about his muscle fiber type distribution and suggested he switch races. The kayaker moved to long-distance competitions and quickly became one of the top racers in the world.

Despite his successful applications of muscle fiber research in track and field and kayaking, soccer vexes Andersen. Soccer coaches all want the fastest athletes, so Andersen wondered how it could be that many of the Danish pros have fewer fast-twitch fibers than an average person on the street. He turned to F.C. Copenhagen's development academy, where he found that the swiftest players are lost to chronic injuries before they ever reach the top level. "The guys that have the very fast muscles can't really tolerate as much training as the others," he says. "The guys with a lot of [fast-twitch fibers] that can contract their muscles very fast have much more risk of a hamstring injury, for instance, than the guys who cannot do the same type of explosive contraction but who never get injured."

The less injury-prone players survived the development years, which is why the Danish elite level ended up skewed toward the slow-twitch. "In American football," Andersen says, "the big fat guy becomes one position and the fast guy becomes a wide receiver and they train differently. But the soccer players are trained all alike. I hear coaches say all the time, 'We can't use him because he's always injured.' If he gets injured all the time, it's probably because we do something wrong to him and we need to change that. We shouldn't lose the fastest players."

Even with all the money and glory available in international soccer, coaches—at least in Denmark—may be losing some of the most fleet players before they ever reach the professional pitch. The same medicine should not be prescribed for every athlete. For some, *less* training is the right medicine.

If innate body type differences that are hidden from our naked eyes, like fiber type proportions, are not accounted for, some athletes are sacrificed to the idea that the same hard training works for everyone. The slow-twitch kayaker who turned from sprints to long-distance might have squandered his career losing shorter races if Andersen had not steered him toward the long-distance races he could win.

In other cases, it is much more obvious how fixed physical traits fit into particular sports amid the rapidly shifting gene pool of competitive athletics.

7

The Big Bang of Body Types

Decades ago, particularly in Europe, local club sports teams supported a large number of regionally competitive, or even semiprofessional, athletes who often made up humanity's elite performers. Until technology tilted the landscape.

Today, literally billions of customers have a ticket to the Olympics, the World Cup, or the Super Bowl with the flick of a remote control. As a result, most sports enthusiasts are now spectators to the elite as opposed to participants in the comparatively quotidian, a huge population of recliner-bound quarterbacks paying to watch a tiny number of real QBs. That scenario creates what economist Robert H. Frank termed a "winner-take-all" market. As the customer base for viewing extraordinary athletic performances expanded, fame and financial rewards slanted toward the slim upper echelon of the performance pyramid. As those rewards have increased and become concentrated at the top level, the performers who win them have gotten faster, stronger, and more skilled.

A group of sports psychologists, particularly acolytes of the strict 10,000-hours school, have argued that improvements in individual sport world records and team sport skill levels have increased so vastly in the last century—faster than evolution could have significantly altered the gene pool—that the improvement must come down solely to increasing amounts of practice. As the rewards for top performers have grown, more athletes have undertaken greater quantities of practice in an attempt to earn them.

A portion of the improvements, though, even in straightforward athletic endeavors, are very clearly the result of technological enhancements. Biomechanical video analysis of legendary sprinter Jesse Owens, for example, has shown that his joints moved as fast in the 1930s as those of Carl Lewis in the 1980s, except Owens ran on cinder tracks that stole far more energy than the synthetic surfaces where Lewis set his records.

But technology is not the only source of improvement that is often overlooked. Undoubtedly, the increasing amount and precision of practice has helped push the frontiers of performance. But the winner-take-all effect, combined with a global marketplace that has allowed many more people to audition for the minuscule number of increasingly lucrative roster spots, has indeed altered the gene pool. Not the gene pool in all of humanity, but certainly the gene pool within elite sports.

In the mid-1990s, Australian sports scientists Kevin Norton and Tim Olds began compiling data on the body types of athletes to see whether there had been significant changes over the twentieth century. The sports science, after all, had changed drastically.

In the late nineteenth century, researchers of the science of body types—known as anthropometry—arrived at conclusions influenced by classical philosophy, like Plato's concept of ideal forms; by art, such as Leonardo da Vinci's *Vitruvian Man*, the famous depiction of a man's body inscribed in a circle and square indicating the ideal human proportions; as well as by racially charged agendas. "There is a perfect form or type of man," reads a late-nineteenth-century article enumerating the characteristics of an athlete, "and the tendency of the race [i.e. the white race] is to attain this type."

At the time, anthropometrists felt that human physique was

distributed along a bell curve, and the peak of the curve—the average—was the perfect form, with everything to the sides deviating by accident or fault. So they asserted that the best athletes would have the most well-rounded, or average, physical builds. Not too tall or too small, neither too skinny nor too bulky, but rather a just-right Goldilocks-porridge version of a man. (And it was only men.) That was the belief for any sport: the average human form would be ideal for all athletic pursuits. This confluence of subjective theory and philosophy dominated the agenda for coaches and physical education instructors in the early twentieth century, and it showed in athletes' bodies. In 1925, an average elite volleyball player and discus thrower were the same size, as were a world-class high jumper and shot putter.

But, as Norton and Olds saw, as winner-take-all markets emerged, the early-twentieth-century paradigm of the singular, perfect athletic body faded in favor of more rare and highly specialized bodies that fit like finches' beaks into their athletic niches. When Norton and Olds plotted the heights and weights of modern world-class high jumpers and shot putters, they saw that the athletes had become stunningly dissimilar. The average elite shot putter is now 2.5 inches taller and 130 pounds heavier than the average international high jumper.

On a height-versus-weight graph, the duo plotted the average physiques of elite athletes in two dozen sports; one data point for the average build of an athlete in each sport in 1925, and another for the average build of an athlete in that same sport seventy years later.

When they connected the dots from 1925 to the present for each sport, a distinct pattern appeared. Early in the twentieth century, the top athletes from every sport clustered around that "average" physique that coaches once favored and were grouped in a relatively tight nucleus on the graph, but they had since blasted apart in all directions. The graph looked like the charts that astronomers constructed to show the movement of galaxies away from one another in our expanding universe. Hence, Norton and Olds called it the Big Bang of body types.

Just as the galaxies are hurtling apart, so are the body types required for success in a given sport speeding away from one another toward their respective highly specialized and lonely corners of the athletic physique universe. Compared with all of humanity, elite distance runners are getting shorter. So are athletes who have to rotate in the air—divers, figure skaters, and gymnasts. In the last thirty years, elite female gymnasts have shrunk from 5'3" on average to 4'9". Simultaneously, volleyball players, rowers, and football players are getting larger. (In most sports, height is prized. At the 1972 and '76 Olympics, women at least 5'11" were 191 times more likely to make an Olympic final than women under five feet.) The world of pro sports has become a laboratory experiment for extreme self-sorting, or artificial selection, as Norton and Olds call it, as opposed to natural selection.

Big Bang data in hand, Norton and Olds devised a measure they called the bivariate overlap zone (BOZ). It gives the probability that a person randomly selected from the general public has a physique that could possibly fit into a given sport at the elite level. Not surprisingly, as winner-take-all markets have driven the Big Bang of body types, the genes required for any given athletic niche have become more rare, and the BOZ for most sports has decreased profoundly. About 28 percent of men now have the height and weight combination that could fit in with professional soccer players; 23 percent with elite sprinters; 15 percent with professional hockey players; and 9.5 percent with Rugby Union forwards.

In the NFL, one extra centimeter of height or 6.5 extra pounds on average translates into about $45,000 of extra income. (Particular professions that require unique physiques have an even more concentrated winner-take-all structure and outdo even professional sports. The BOZ for regional catwalk models is less than 8 percent, dropping to 5 percent for international models, and to just 0.5 percent for supermodels.)

And the Big Bang of body types goes down to the body-part level as well. While tall athletes have grown taller at a much faster rate than humanity as a whole, and small athletes have shrunk relatively smaller, athletes in certain sports have increasingly needed extremely specialized body traits. Measurements of elite Croatian water polo players from 1980 to 1998 show that over two decades the players' arm lengths increased more than an inch, five times as much as those of the Croatian population during the same period. As performance requirements become stricter, only the athletes with the necessary physical structure consistently make the grade at the elite level. The shorter-armed athletes are more often weeded out.

In addition to having longer arms overall, the bone proportions in the arms of top water polo players have changed. Elite players now have longer lower arms compared with their total arm length than do normal people, giving them a more efficient throwing whip. The same is true of athletes who need long levers for powerful, repetitive strokes, like canoeists and kayakers. Conversely, elite weight lifters have increasingly shorter arms—and particularly shorter forearms—relative to their height than normal people, giving them a substantial leverage advantage for heaving weights overhead. One of the many failings of the NFL combine that tests prospective draft picks in physical measures is that arm length is not taken into account in the measure of strength. Bench press is much easier for men with shorter arms, but longer arms are better for everything on the actual football field. So a player who is drafted high because of his bench press strength may actually be getting a boost from the undesirable physical characteristic of short arms.

Top athletes in jumping sports—basketball, volleyball—now have short torsos and comparatively long legs, better for accelerating the lower limbs to get a more powerful liftoff. Professional boxers come in an array of shapes and sizes, but many have the combination of long arms and short legs, giving greater reach but a lower and more stable center of gravity.

The height of a sprinter is often critical to his best event. The world's top competitors in the 60-meter sprint are almost always shorter than those in the 100-, 200-, and 400-meter sprints, because shorter legs and lower mass are advantageous for acceleration. (Short legs have a lower moment of inertia, which essentially means less resistance to starting to move.) Sprinters hit the highest top speeds in the 100- and 200-meter races, but the 60-meter race has a proportionally longer acceleration period. Perhaps the advantage of shortness for acceleration explains why NFL running backs and cornerbacks, who make their livings starting and stopping as quickly as possible, have gotten shorter on average over the last forty years, even while humanity has grown taller.

On occasion, technique changes in sports have changed the advantaged body types almost overnight. In 1968, Dick Fosbury unveiled his "Fosbury flop" method of high jump, which gives an advantage to athletes who have a high center of gravity. In just eight years after Fosbury's innovation, the average height of elite high jumpers increased four inches.*

In other cases, body types have more nuanced effects. While smallness is generally a boon for endurance runners, Paula Radcliffe, the world record holder in the women's marathon, at 5'8" is literally head and shoulders above most of her world-class competitors. It didn't keep the iconically tough Brit from winning eight marathons in the prime of her career, 2002 to 2008. But Radcliffe's size may have helped confine most of her victories to autumn. One reason that marathon runners tend to be diminutive is because small humans have a larger skin surface area compared with the volume of their body. The greater one's surface area compared with volume, the better the human radiator and the more quickly the body unloads heat. (Hence,

---

*The global search for increasingly suitable athletic bodies has been wildly successful in just about every sport. For centuries, Japanese competitors dominated Sumo wrestling because, well, only Japanese people were competing. From the seventeenth century until 1990, only Japan-born wrestlers had attained the top rank of Yokozuna. But in the global sports marketplace, athletes from countries with generally larger residents have infiltrated Sumo in a big way. To the dismay of some Sumo traditionalists, five of the last seven Yokozuna have been Mongolian or Hawaii-born Americans.

short, skinny people get cold more easily than tall, hefty people.) Heat dissipation is critical for endurance performance, because the central nervous system forces a slowdown or complete stop of effort when the body's core temperature passes about 104 degrees.*

While Radcliffe in her prime was unbeatable on autumn mornings when races were held in cool temperatures, she was feckless in summer heat. At the Athens Olympics in 2004, when the marathon was held in 95-degree heat, despite having by far the fastest time coming into the race she was unable to finish and crumpled in a heap by the side of the road. The woman who won the race was 4'11". At the Beijing Olympic marathon in 2008, the temperature was 80 degrees and humid and Radcliffe finished a distant twenty-third. From 2002 to 2008, Radcliffe was 8-0 in marathons contested in cool or temperate conditions, and 0-2 and never even in contention in the sweltering summer Olympic races.

Collecting data for the most famous study of athletic body types ever conducted took an international research team a full year and included 1,265 athletes who competed at the 1968 Mexico City Olympics, representing every sport (except equestrian) and 92 different countries. It took six more years for the results to be compiled and published in a 236-page book. Half the book is simply tables of body measurements. Even without text, they convey an obvious message: in most Olympic sports, athletes are generally more physically similar to one another than I am to my own brother.

Within track and field, most of the athletes could be pinned to an

---

*One reason why amphetamines are so good, albeit illegal, for enhancing endurance performance is that they appear to remove the brain's inhibition from overheating, allowing an athlete to continue beyond 104 degrees. Great for performance, but it has also led to heat stroke deaths during competition. In 2009, a Kentucky high school football coach was tried for murder when one of his players collapsed and died in a practice in extreme heat. The coach was acquitted, and it was revealed that the player was taking prescribed amphetamines to treat ADHD.

event simply by their body measurements. The men and women who raced the 400- and 800-meters or the high hurdles were the tallest of the runners—no surprise, given that the goal in hurdling is to clear the barriers with as little movement of the center of gravity as possible—while the marathoners were the shortest. No surprise there, either. But the similarities extended to less obvious physical traits of the skeleton.

Athletes in a sport or event tended to be similar in height and weight—and often different from a control population of nonathletes—and also with respect to the breadth of their pelvic bone and the skeletal structure of their shoulders.

Nonathlete women who were measured as a control group for the study had, of course, wider pelvic bones than nonathlete men. But female swimmers had more narrow pelvic bones than the normal, control population of men. And female divers had more narrow pelvic bones than the female swimmers. And female sprinters more narrow than the female divers. (Slim hips make for efficient running.) And female gymnasts had slimmer hips still.

Female sprinters had much longer legs than the control population of women, and about as long as the control men. Male sprinters were around two inches taller than the control men, and 100 percent of that was in their legs, such that when they were seated the sprinters were the same height as the control men.

The male swimmers were, on average, more than 1.5 inches taller than the sprinters, but nonetheless had legs that were a half-inch shorter. Longer trunks and shorter legs make for greater surface area in contact with the water, the equivalent of a longer hull on a canoe, a boon for moving along the water at high speed. Michael Phelps, at 6'4", reportedly buys pants with a 32-inch inseam, shorter than those worn by Hicham El Guerrouj, the Moroccan runner who is 5'9" and holds the world record in the mile. (Like other top swimmers, Phelps also has long arms and large hands and feet. That elongated body type can be indicative of a dangerous illness called Marfan syndrome.

According to Phelps's autobiography, *Beneath the Surface,* his unusual proportions led him to get checked annually for Marfan.)*

The more that elite sports markets have shifted from participatory affairs to events for bulging masses of spectators, the more rare the bodies required for success have become, and the greater the money needed to attract those rare bodies to a particular sport. In 1975, athletes in the major American sports averaged roughly five times the median salary for an American man. Today, the average salaries in those sports are between about forty and one hundred times the median full-time salary. In order to match a single year's salary of the highest-paid athletes, an American man making the country's median annual income for a full-time job would have to work for five hundred years.

Genes affect body weight. The GIANT (Genetic Investigation of ANthropometric Traits) Consortium study of 100,000 adults found six DNA variants that influence heft. The FTO gene, in itself, accounts for several pounds in studies, possibly by influencing one's taste for fatty foods. But, as anyone who has ever gorged himself on Thanksgiving dinner and then hopped on a scale can attest, weight is substantially affected by lifestyle.

Fat is the tissue in the body that is most responsive to training and diet. (And weight is extremely responsive to certain drugs. When Norton and Olds examined the inflating size of NFL defensive tackles, they found an eye-catching acceleration in size in the late 1960s and early '70s, when steroids began to proliferate in football. From the 1940s to the 1990s, the body mass index of an NFL defensive tackle rose from 30 to 36. For a 6'2" tackle, that's a rise in weight from 234 pounds to 280 pounds.)

---

*In sports like swimming, kayaking, and lacrosse, athletes tend to have a very high "brachial index." That is, the forearm is relatively long compared to the upper arm, which makes the arm better suited to propulsion. Weight lifters and wrestlers, who need stability and strength, have very low brachial indices.

Clearly, the FTO gene has been around since long before the recent obesity epidemic in the industrialized world. More genes that influence weight are sure to be found—studies of twins and adopted children suggest there are more out there—and the complex interplay of genetics, lifestyle, and weight are only beginning to be illuminated. Even combined, all of the DNA variants that the GIANT Consortium identified accounted for only a small fraction of bulk. (Based on my DNA analysis, I am entitled to attribute just 8.5 of my 150 pounds to those genes.)

And just as an individual's proportion of fast- and slow-twitch muscle fibers influences his muscle growth potential, it also influences his fat-burning capacity. Researchers in the United States and Finland have independently shown that, while adults with a high proportion of fast-twitch fibers can pack on muscle, they have a more difficult time losing fat. Fat is primarily burned as part of the energy-making process that occurs in slow-twitch muscle fibers. The fewer slow-twitch muscle fibers an individual has, the lower his capacity to burn fat—one possible reason that sprint and power athletes tend to be stockier than endurance athletes, even before and after their competitive years.

And while it is obvious that diet and training can dramatically alter an athlete's build, there are limits. Limits delineated by an individual's skeleton.

Francis Holway, an exercise and nutrition researcher in Buenos Aires, has been obsessed with the limits of body types since childhood. His first inspiration was the story of Tarzan. He was fascinated by how the son of a British lord adopted by apes and transplanted to a jungle environment could develop the rhino-wrestling physique and vine-swinging skills to thrive. Holway's first experiments, at the age of seven, came when he would gulp down spoonfuls of oatmeal and then flex his biceps right after the meal to see if they had grown.

As a kid, he first thought that the sport shaped the body; that basketball players would grow tall from playing and weight lifters would become squat from squatting. To some degree, the research he has conducted as an adult has borne out similarly startling phenomena. Holway measured the forearms of a group of tennis players ranked in the top twenty in the world and found that their racket arms grew slightly differently from their nonracket arms. The racket-side forearm bones of the players grew around a quarter-inch longer than the forearm bone of the nonracket arm. And the elbow joint widened a centimeter. Like muscle, bone responds to exercise. Even nonathletes tend to have more bone in the arm they write with simply because they use it more, so the bone becomes stronger and capable of supporting more muscle. "It's just amazing how the bones can adapt to repeated stress," Holway says. Those tennis pros literally served and volleyed their ways to longer forearms. And yet, this malleability is limited.

Libby Cowgill, an anthropologist at the University of Missouri, has studied skeletons from around the world in an effort to determine whether certain populations have built strong skeletons through childhood activity or whether they are simply born with robust skeletal scaffoldings capable of supporting mounds of muscle. "We can see differences in the strength of bones in different populations already at one year of age," Cowgill says. "What I've found indicates that these differences are just there. They are exacerbated over the course of growth based on what you're doing, but it looks like people are born with genetic propensities to be strong or to be weak."

In one study, she compared the skeletons of Mistihalj people—a group of medieval Yugoslavian herders—to the skeletons of kids from 1950s Denver. "The herders' kids are the biggest, buffest kids I've ever seen," she says. "Based on data of modern American children, we're just puny in terms of the amount of bone we have." But might a strict childhood training program be able to transform any American tot into a mighty medieval herder? "There's a lot you can do with activity, and especially starting it earlier," Cowgill says. "But it's looking more and more like there's a genetic component as well."

The skeleton you are bequeathed has a lot to do with whether you will ever be able to make the weight required for a particular sport. Holway compares the skeleton to an empty bookcase. One bookcase that is four inches wider than another will weigh only slightly more. But fill both cases with books and suddenly the little bit of extra width on the broader bookcase translates to a considerable amount of weight. Such is the case with the human skeleton. In measurements of thousands of elite athletes from soccer to weight lifting, wrestling, boxing, judo, rugby, and more, Holway has found that each kilogram (2.2 pounds) of bone supports a maximum of five kilograms (11 pounds) of muscle. Five-to-one, then, is a general limit of the human muscle bookcase.*

"We've had people come in for consultation and they want to increase their muscle mass for aesthetic reasons," Holway says. "We measure them, and if they're close to five-to-one we ask them how long they've been at this same level of development or strength. They'll say for the last five years or seven years, and they haven't been able to surpass it." Holway experimented on himself, spending years in heavy weight training with a diet high in protein and supplemented by creatine. But as he closed in on five-to-one, inhaling more steaks and shakes only added fat, not muscle.

Male Olympic strength athletes whom Holway has measured, like discus throwers and shot putters, have skeletons that are only about 6.5 pounds heavier than those of average men, but that translates to more than 30 pounds of extra muscle that they can carry with the proper training. Holway uses his measurements to help tailor athletes' training. In the shot put, for example, an athlete needn't move himself very far, so even adding extra fat might be worthwhile, since the athlete needs to pack on bulk to become relatively more massive than

---

*The limit Holway has documented for women is closer to 4.2 to 1. And both limits are sans steroids. Athletes on steroids have been able to surpass the 5 to 1 upper bound.

the object being thrown. But in javelin, where the athlete needs to both run fast and throw hard, he should be wary of trying to add weight beyond the five-to-one ratio, as it will likely be fat. Or consider a Sumo wrestler, or an offensive lineman in football who simply wants to be difficult for his opponent to move. He might do well to add extra fat. Offensive linemen are incredibly strong, but they most certainly are not ripped.

Again, when innate biological differences are taken into account, it becomes clear that successful training plans are those tailored to the individual's physiology. As Dr. J. M. Tanner, an eminent growth expert (and world-class hurdler), wrote in *Fetus into Man*: "Everyone has a different genotype. Therefore, for optimal development, everyone should have a different environment."

Heaving sports performance to untouched heights requires both specialized training and specialized bodies to be trained.

Today, the expanding universe of athletic body types is slowing down. Much of the self-sorting, or artificial selection, is finished. The tall athletes are no longer getting taller compared with the rest of humanity at the rate they were two decades ago, nor the small smaller. And the march of constantly shattered world records is slowing right along with it.

Over most of the twentieth century, the adage "records were made to be broken" rang chronically true. But athletic records in most, but certainly not all, events that have high historical participation are now inching forward—if, that is, they are moving forward at all. The coveted world records in the men's mile and 1,500-meters (the race close to the mile that is run outside the United States) were broken collectively about eight times per decade from the 1950s to 2000, but not at all since. Other records have continued to creep down, but usually by small margins. It will be intriguing to see whether the financial success of Usain Bolt, who dropped records by rather large margins, draws more athletes with his unusual combination of explosiveness and height away from other sports to sprinting.

"There are still some unexploited parts of the world, but we've reached much of the global market," says Tim Olds, one of the Big Bang of body types scientists. "We're getting closer to reaching the limit of our source populations for bodies. Population progression is slowing globally, so we're going to see slowing growth in both body size and body shapes, and in records as well." Just as exploring the earth must once have seemed like an endless endeavor for adventurers, perhaps the era of constant record shattering is largely in the past, and the future will be one of baby steps forward.

As the expanding universe of sports physiques has sped outward, finding those increasingly rare bodies has fostered an increasingly extensive, and expensive, global talent search.

In that endeavor, no league has been more successful than the National Basketball Association.

8

The Vitruvian NBA Player

ong before he became a pop-culture trope, before he dated Ma-
donna or married Carmen Electra, or married *himself* as a publicity
stunt; before he posed for the cover of *Sports Illustrated* with fire-engine-
red hair, wearing a metal-studded choker and a smug look and holding
a blue parrot; before he announced he would start a topless women's
basketball league, and way before he hung out with North Korean
leader Kim Jong-un, Dennis Rodman was just an insecure little boy.

Every night before he fell asleep as a child in the Oak Cliff housing
projects in Dallas, he would lie awake and think: "There's something
big out there waiting for Dennis Rodman." Little did he know at the
time, the *something big* would be himself.

Back then, Rodman's sisters were the basketball stars. Both would
become college All-Americans, while Dennis, the family runt, was
short and awkward and struggled to sink a layup. He warmed the
bench for a half season of high school basketball and then quit. He
was 5'9" when he graduated, and endured taunting by his friends
when he tagged along with his bigger, younger, more athletic sisters.

After graduation, Rodman took a job on the graveyard shift sweeping
floors at the Dallas/Fort Worth International Airport. One night, he
stuck a broom through the safety gate of a shuttered airport gift shop
and fished out a few dozen watches that he distributed among his friends.

He got caught. Rodman didn't last long in that job. But his *something big* had already started to happen. In the two years since high school, Rodman had grown like kelp. He was working part-time scrubbing cars at an Oldsmobile dealership for $3.50 an hour when he topped out at 6'8".

So Rodman started to play basketball and found that he was suddenly *less* gawky despite being taller and more muscular. He caught on to the game so quickly it was as if the basketball fairy had left hoop skills under his pillow one night. In his words: "It was like I had a new body that knew how to do all this shit the old one didn't."

A family friend convinced Rodman to try out for a local community college team. He played for a while, but dropped out with academic problems. The following year, 1983, he took a basketball scholarship to Southeastern Oklahoma State, a little-known NAIA school. He dominated there for three years, averaging 25.7 points and an otherworldly 15.7 rebounds per game. The rest is hardwood history. Rodman was drafted into the NBA and in fourteen years won five championships, was twice named Defensive Player of the Year, and became the greatest rebounder in NBA history. In 2011, the man who played hardly any organized basketball before he was twenty-one was inducted into the Basketball Hall of Fame.

Only marginally less inevitable than death and taxes during the 1990s was the Chicago Bulls winning the NBA championship.

Dynastic dominance came on the backs of three future Hall of Famers, and three nick-of-time growth spurts. Before the triune pillars of the Bulls dynasty had their height, their skills alone did not elevate them above the crowd.

There was Rodman, of course. Then there was Scottie Pippen, who had a similar story. He was 6'1" when he graduated from high school and started out as the team manager at the University of Central Arkansas. He sprouted to 6'3" by the end of his first year and started playing for the team. By the end of the following summer, Pippen was 6'5". By his junior season, Pippen was 6'7" and NBA scouts began to swarm the stands to watch unheralded Central Arkansas. Years later, he was selected as one of the fifty greatest players in NBA history and was inducted into the Hall of Fame one year before Rodman.

Michael Jordan didn't cut it quite so close. Jordan was already a good basketball player in high school—he started dunking as a 5'8" freshman—but he comes from a comparatively diminutive family and was already anomalous at six feet as a high school sophomore. As a high school junior, Jordan was being evaluated by college scouts, but looked like a better fit for a small school. By Jordan's own reckoning, his 5'7" brother Larry was as athletic as he and held the upper hand in their backyard games, until Michael's kelp phase. He grew six inches late in high school and dropped baseball to focus on basketball. He earned a scholarship to powerhouse North Carolina. The rest of the story hardly needs telling.

Rodman, Pippen, and Jordan formed the nucleus of a Bulls team that went 72-10 in the 1995–96 season, a feat unequaled before or since, and their biographies are testaments to the primacy of height.

That isn't to suggest that being 6'6" or 6'8" automatically makes a professional basketball player, much less a Hall of Famer. As ESPN personality Colin Cowherd said on his radio show, "Talent doesn't fall out of the womb . . . there are a million guys in America who are six-foot-eight who aren't in the NBA." But then, that's not right either.

Based on data from the United States Census Bureau and the National Center for Health Statistics, there are likely fewer than twenty thousand American men between the ages of twenty and forty who are at least 6'8". So a Dennis Rodman or a LeBron James is not one in a million—compared to men of equal height—but rather one in a pool the size of Rolla, Missouri.

Height is an incredibly narrowly constrained trait among humans. Fully 68 percent of American men are in just the six-inch range from 5'7" to 6'1". The bell curve of adult height is a Himalayan slope that falls off precipitously on either side of the mean. A mere 5 percent of American men are 6'3" or taller, while the average height of an NBA player perennially hovers around 6'7". Suffice it to say that there is startlingly little overlap—far less than Cowherd suggested—between the heights of humanity and those of NBA players.

While inhabitants of the industrialized world grew taller over much of the twentieth century at a rate of about one centimeter per decade—at least partly because of increased protein intake and the decline of growth-stunting childhood infections, and perhaps because people are mixing genes more widely, with "tall" genes dominating "short" genes—NBA players have been growing at more than four times that rate, and the tallest of the tall NBA players at ten times that rate.

In *Outliers*, Malcolm Gladwell makes a point about height in basketball by comparing it to IQ. There is a threshold, he writes, above which more does not really matter. Above an IQ of 120—which already eliminates most of humanity—he argues, one is already smart enough to consider the most difficult intellectual problems, and more IQ does not translate into real-world success. In basketball, he adds, "it's probably better to be six two than six one . . . But past a certain point, height stops mattering so much." But the "threshold hypothesis" of IQ is not supported by the work of scientists who specialize in that field, nor is the threshold hypothesis of NBA height supported by player data.

Based on data from the NBA and NBA predraft combines (using only true, shoes-off measurements of players), the Census Bureau, and the Centers for Disease Control's National Center for Health Statistics, there is such a premium on extra height in the NBA that the probability of an American man between the ages of twenty and forty being a current NBA player rises nearly a full order of magnitude with every two-inch increase in height starting at six feet. For a man between six feet and 6'2", the chance of his currently being in the NBA is five in a million. At 6'2" to 6'4", that increases to twenty in a million. For a man between 6'10" and seven feet tall, it rises to thirty-two thousand in a million, or 3.2 percent. An American man who is seven feet tall is such a rarity that the CDC does not even list a height percentile at that stature. But the NBA measurements combined with the curve formed by the CDC's data suggest that of American men ages twenty to forty who stand seven feet tall, a startling 17 percent of them are in the NBA *right now*.* Find six honest seven-footers, and one will be in the NBA.

Kevin Norton and Timothy Olds, the Big Bang of body types scientists, charted the increase of seven-foot players in the NBA from 1946 to 1998 and found that the proportion of seven-foot NBA players rose slowly but steadily for thirty-five years, from zero in 1946 to about 5 percent of all players in the early 1980s, just before the winner-take-all basketball market kicked into hyperspeed.

In 1983, the NBA struck a groundbreaking collective bargaining agreement with players that made the athletes partners in the league, entitled to money from licensing agreements, ticket sales, and television contracts. The following year, a rookie Michael Jordan signed a comparably trailblazing contract with Nike that gave him royalties from the sales of sneakers bearing his name.

Suddenly, the earning potential of professional basketball players shot through the arena roof, and pretty much anyone who *could* play in the NBA wanted to. At the same time, NBA teams began scouring the globe for giants. In just three years following the new labor agreement, the proportion of seven-footers in the NBA more than doubled, reaching 11 percent, where it has essentially remained ever since. "What it means is that basically everyone in the world who is seven foot tall and can play basketball is part of the game," Olds says. "We've kind of reached a population limit."

Reaching it required increasing globalization of the game. The average height of American players in the NBA is about 6'6½", while the average height of foreign players is nearly 6'9". A great many of the foreign players in the NBA are there, it seems, because teams ran low on sufficiently tall players at home. Perhaps it's no surprise, then, that the

---

*Many of the men who NBA rosters claim are seven feet tall prove to be an inch or even two inches shorter when measured at the combine with their shoes off. Shaquille O'Neal, however, is a true 7'1" with his shoes off.

non-U.S. countries with stable representation in the NBA—Croatia, Serbia, Lithuania—are among the tallest in the world. Because height is a "normally distributed" human trait (i.e., a bell curve), a tiny difference in the average height in a country means a big difference in the number of people at the far extremes, like seven-footers.

In terms of freakish height, the Women's National Basketball Association lags far behind the men's league. The average height of a WNBA player is between 5'11" and six feet, not as relatively tall compared with an average woman as an NBA player with an average man. The average WNBA player is only about 10 percent taller than the average American woman, compared with the average NBA player who is closer to 15 percent taller than the average man.

Perhaps it will just take time for more tall women to gravitate to the game. Or perhaps it will take a stronger winner-take-all market. WNBA players make just tens of thousands of dollars per year, while the average NBA player rakes in more than $5 million a year. It is easy to see why many women with the athletic gift of height might be inclined toward other sports that hold more lucrative opportunities for them, like tennis. As rackets have grown lighter and serves more important in tennis, players have gotten taller. At the time of this writing, the top three female tennis players in the world have an average height of 5'11⅔", nearly identical to the average height in the WNBA.

None of this is to say that shorter men and women can't succeed in basketball. NBA players like Muggsy Bogues (5'3"), Nate Robinson (a shade under 5'8"), and Spud Webb (5'7", with thick socks) all thrived in the land of giants. But not without abilities that compensated for their stature. Robinson and Webb, two of the shortest players in NBA history, both won the Slam Dunk Contest. Bogues claimed an astonishing forty-four-inch vertical leap, but his tiny hands made it difficult to palm a basketball, so he was content to dunk volleyballs in practice.

Short people generally don't make the NBA unless they have extraordinarily anomalous jumping ability. Not necessarily like Bogues, Robinson, and Webb, but consider the all-time grand total number of men drafted into the NBA who were unable to get high enough to grab the rim when tested at a predraft combine: zero. But there's something else that helps wee NBA ballers play large.

Leonardo da Vinci's *Vitruvian Man* has an arm span equal to his height. So do I. So, probably, do you, or very nearly so. Nate Robinson, on the other hand, is 5'7¾" and his arm span is 6'1". He is, effectively, not as short as he is. Actually, almost none of the players in the NBA are as short as they seem, including the ridiculously tall ones.

The average arm-span-to-height ratio of an NBA player is 1.063. (For medical context, a ratio of greater than 1.05 is one of the traditional diagnostic criteria for Marfan syndrome, the disorder of the body's connective tissues that results in elongated limbs.) An average-height NBA player, one who is about 6'7", has a wingspan of seven feet. To fit the Vitruvian NBA player, Leonardo would have needed a rectangle and an ellipse, not his tidy square and circle.

NBA players who are labeled as "undersized" for the position they play based on stature generally have the extra arm span to make up for it. Elton Brand, the first pick of the 1999 NBA Draft, at 6'8¼" is unremarkable for a power forward. But Brand is actually a giant among power forwards if you consider his 7'5½" of reach. John Wall, the point guard who was the first pick in the 2010 draft, is only 6'2¾" with his shoes off, but has 6'9¼" worth of reach. When the Miami Heat assembled its ballyhooed Big Three—Chris Bosh, LeBron James, and Dwyane Wade—before the 2010–11 season, the team was enlisting 19'9¼" of height, but 21'2½" of wingspan. And it's no coincidence.

Based on statistics for players who were on NBA rosters at the beginning of the 2010–11 season, a player's wingspan influences a number of key statistics. An NBA general manager who wants to increase his team's blocked shots would be better off signing a player with an extra inch of arm than an inch of height. The New Orleans Pelicans' Anthony Davis, the shot-swatting first pick of the 2012 draft, is 6'9¼" with a 7'5½" wingspan. A player with Davis's build will be predicted to get ten more blocks per season than a 7'1" giant who plays an equal number of minutes but has arms that match his height. If the GM wanted offensive rebounds, he would do equally well to sign a player with an extra inch of reach as an extra inch of height. And while height is a slightly better predictor of defensive rebounds than is wingspan, both are important and together account for half of an NBA player's defensive boards, without even considering characteristics like jumping ability, weight, position, or general rebounding skill.

Stat-savvy general managers have no doubt noticed. Daryl Morey, the MIT-educated GM of the Houston Rockets, renowned for his Moneyball approach to basketball, has drafted several of the most superficially undersized players in the NBA. (Morey would not comment on whether the Rockets strategically targeted high wingspan-to-height-ratio players in the draft.) Three seasons ago, the Rockets used the shortest starting center in NBA history, Chuck Hayes, who is just 6'5½". Fortunately, his arms are 6'10".

The bottom line is that not only are NBA players outlandishly tall, they are also preposterously long, even relative to their stature. And when an NBA player does not have the height required to fit into his slot in the athletic body types universe, he nearly always has the arm span to make up for it. In the post–Big Bang of body types era, whether with height or reach, almost no player makes the NBA without a functional size that is typical for his position and often on the fringe of humanity. Only two players from a 2010–11 NBA roster with available official measurements have arms shorter than their height. One is J. J. Redick, the Milwaukee Bucks guard who is 6'4" with a 6'3¼" arm span, downright *Tyrannosaurus rex*-ian in the NBA.* The other is now-retired Rockets center Yao Ming. But at a height just over 7'5", Yao, whose gargantuan parents were brought together for breeding purposes by the Chinese basketball federation, fit into his niche just fine.

---

*Accomplished boxers often have long arms as well, but the trend is nowhere near as ubiquitous as in the NBA, even among the greatest heavyweights. Rocky Marciano was the J. J. Redick of his boxing era, standing 5'11" with a reported 5'7" reach. Meanwhile, Sonny Liston was 6'0" with a reach of 7'0".

Repeatedly, studies of families and twins find the heritability of height to be about 80 percent. That means that 80 percent of the difference in height between people in the group that is being studied is attributable to genetics, and around 20 percent to the environment. (In nonindustrialized societies, the heritability of height is lower, as many citizens, like plants in poor soil, are prevented by nutritional deficiencies or infections from reaching their genetic height potential.) So if the tallest 5 percent of citizens in a given population are a foot taller than the shortest 5 percent, genetics will account for about ten inches of the disparity.

For much of the twentieth century, denizens of industrialized societies were growing taller at a rate of about one centimeter per decade. In the seventeenth century, the average Frenchman was 5'4", which is now the average for an American woman. The first generation of Japanese born to immigrant parents in America, known as the Nisei, famously towered over their parents.

In the 1960s, growth expert J. M. Tanner examined a set of identical twins that suggested the range of height variability caused by the environment. The identical boys were separated at birth, one brother raised in a nurturing household, and the other reared by a sadistic relative who kept him locked in a darkened room and made him plead for sips of water. In adulthood, the brother from the nurturing household was three inches taller than his identical twin, but many of their body proportions were similar. "The genetic control of shape is more rigorous than that of size," Tanner wrote in *Fetus into Man*. The smaller brother was an abuse-shrunken version of the bigger brother.

Little is known about the actual genes that influence height, however, because the genetics of even outwardly simple traits tend to be very complicated. A 2010 study in *Nature Genetics* needed 3,925 subjects and 294,831 single nucleotide polymorphisms—spots of DNA where a single letter can vary between people—to account for just 45 percent of the variance in height between adults, and that's the best any study has done. Finding all the height genes will take much larger and more complex studies than scientists presumed a decade ago.

Though the genes are difficult to pinpoint, the genetically programmed nature of height is obvious from studies of identical twins. Due to distinct intrauterine conditions, identical twins are often *less* similar in birth size than fraternal twins. And yet, after birth, the smaller twin of an identical duo quickly catches up with the bigger twin and they will be nearly or exactly the same height as adults. Similarly, female gymnasts delay their growth spurt with furious training, but that does not diminish their ultimate adult height. The genetic programming is also evident in the rate at which children grow. In World Wars I and II, European children were exposed to brief periods of famine during which their growth ground almost to a halt. When food again became plentiful, their bodies put the growth pedal to the metal such that adult height was not curtailed. "The undernourished child slows down and waits for better times," Tanner wrote. "All young animals have the capacity to do this. . . . Man did not evolve in the supermarket society of today."

The permutations of size-determining interactions between nature and nurture are fathomless. Consider that children grow more quickly in spring and summer than in fall and winter, and that this is apparently due to sunlight signals that enter through the eyeballs, since the growth of totally blind children consists of similar fluctuations but are not synchronized with the seasons.

The height that inhabitants of urban societies gained over the twentieth century came principally from increased leg length. Legs got longer faster than torsos. In developing countries that have gaping nutritional and infection-prevention disparities between the middle class and poor, the difference in height between the comfortable and the afflicted is all in the legs.

Japan displayed a startling growth trend during its "economic miracle" period following World War II. From 1957 to 1977, the average height of a Japanese man increased by 1.7 inches, and of a woman by an inch. By 1980, the height of Japanese people in Japan had caught up with the height of Japanese people in America. Amazingly, the entire height increase was accounted for by increased leg length. Modern Japanese people are still short compared with Europeans, but not as short as they once were. And they now have more similar proportions.

There are, however, certain body type differences that have persisted over time and that have attracted the interest of sports anthropometrists. Every study that has examined race differences in body types has documented a disparity between black and white people that remains whether they reside in Africa, Europe, or the Americas. For any given sitting height—that is, the height of one's head when one is sitting in a chair—Africans or African Americans have longer legs than Europeans. For a sitting height of two feet, an African American boy will tend to have legs that are 2.4 inches longer than a European boy's. Legs make up a greater proportion of the body in an individual of recent African origin.* And this holds for elite athletes.

Studies of Olympic athletes are uniformly consistent in finding that Africans and African Americans and African Canadians and Afro-Caribbeans have a more "linear" build than their competitors of Asian and European descent. That is, they tend to have longer legs and more narrow pelvic breadth.

In their summary of the measurements of 1,265 Olympians from the 1968 Olympics in Mexico City, the scientists state that the successful body types within a sport are much more similar than body types between sports, regardless of ethnicity, but that "the most persistent of these differences" within sports are the narrow hip breadths and longer arms and legs of athletes with recent African ancestry. "They appear in virtually all the events," the researchers write.

---

*Because people with recent African ancestry tend to have longer limbs, the traditional diagnostic criteria for Marfan syndrome have been updated such that they are different for African Americans and white Americans. For African Americans, a trunk-to-legs ratio of less than 0.87 may be indicative of Marfan, whereas in white Americans the diagnostic ratio is 0.92.

Modern scientists who have measured athletes mention in their writing, sometimes reluctantly, that these body type differences influence athletic performance. The scientists are often careful to point out that a particular body type is not *better* overall, but that it may fit more readily into one sports niche than another. "This pattern may, in part, explain the tendency for the linear and relatively long-limbed east Africans to excel in endurance events while the short-limbed eastern Europeans and Asians have a long history of success in weight lifting and gymnastics," write Norton and Olds, the Big Bang of body types gurus, in their textbook *Anthropometrica*.

The limb-length difference manifests in NBA data as well.* In NBA predraft measurements for active players, the average white American NBA player was 6'7½" with a wingspan of 6'10". The average African American NBA player was 6'5½" with a 6'11" wingspan; shorter but longer. Both white and black players in the NBA have wingspan-to-height ratios much greater than the population average, but there's a sizable gap between white and black players. The average ratio for a white American NBA player is 1.035, and for an African American NBA player 1.071. Still, there is wide variation among players within a given ethnicity. Two white players, Coby Karl (height: 6'3½", wingspan: 6'11") and Cole Aldrich (height: 6'9", wingspan: 7'4¾"), for example, have wingspan-to-height ratios approaching 1.10, but they are significant outliers compared with the other white players in the NBA. No other white players are even close, whereas a number of black players have larger ratios. When I showed this data to a scientist who studies athletes' bodies, he responded: "So maybe it's not so much that white men can't jump. White men just can't reach high."†

In a sense, this is last millennium's news to scientists who have

---

*Data on the ethnicity of NBA players was matched with data generously shared by Brigham Young University economist Joseph Price, who has done fascinating work analyzing racial bias among NBA referees as it pertains to foul calls.

†The measurements from the NBA combine used for this chapter speak only to a very specialized sample of athletes, but, based on that data, the average standing vertical jump for a white NBA player at the combine was 27.29 inches, and for a black player 29.64 inches.

been studying body forms. In 1877, American zoologist Joel Asaph Allen published a seminal paper in which he noted that the extremities of animals get longer and thinner as one travels closer to the equator. African elephants can be distinguished from Asian elephants by their sail-like floppy ears. This is because the ears, like your skin, act as a radiator to release heat. The greater the surface area of the radiator compared with its volume, the more quickly heat is released. The African elephants, having evolved closer to the equator, have developed larger ears for cooling purposes. "Allen's rule," that animals from warmer climates tend to have longer limbs, has been extended to humans by a veritable filing cabinet full of studies.

A 1998 analysis of hundreds of studies of native populations from around the world found that the higher the average annual temperature of a geographic region, the proportionally longer the legs of the people whose ancestors had historically resided there. Men and women from dozens of native populations on every inhabited continent were included, and when it came to leg length, they grouped by geography. Low-latitude Africans and Australian Aborigines had the proportionally longest legs and shortest torsos. So this is not strictly about ethnicity so much as geography. Or latitude and climate, to be more precise. Africans with ancestry in southern regions of the continent, farther from the equator, do not necessarily have especially long limbs. But whether an African person in the study was from a population in Nigeria or from a genetically and physically distinct population in Ethiopia, so long as he was from low latitude his legs were likely longer than those of a height-matched European. And certainly longer than those of an Inuit from northern Canada, as Inuit tend to be short and stocky with compact limbs and a wide pelvis.*

In the nineteenth century, Allen surmised that the long limbs of low-latitude animals were a *direct* result of a warm climate. In other

---

*It's important to remember that these are average statements. We can all agree, for example, that men are taller than women on average. And yet, there is enough individual variation that it is easy to find a woman who is taller than many men.

words, he guessed that if a baby African elephant were adopted by Asian elephant parents and raised at high latitude in Asia, it would have the same smallish ears as Asian elephants. On that point, he was mistaken. Comparisons of human descendants of equatorial Africans and of Europeans who now live in the same country, like England or the United States, show that the limb differences remain. The effect of climate on extremities is therefore primarily through genetic selection over generations. Ancestral humans with shorter limbs had a greater chance of surviving and reproducing in cold northern latitudes because they retained more heat.

In 2010, a racially diverse research team from Duke and Howard universities confronted the issue of body types as it pertains to ancestry and sports performance. The scientists did a backbend to avoid racial stereotyping. "Our study does not advance the notion of race," they wrote. In a press release accompanying the study, Edward Jones, a black member of the research team, emphasized that access to sports facilities is critical for athletic development and that while growing up in South Carolina he was discouraged from swimming. Nonetheless, the researchers reported that, compared with white adults of a given height, black adults have a center of mass—approximately the belly button—that is about 3 percent higher. They used engineering models of bodies moving through fluids—air or water—to determine that the 3 percent difference translates into a 1.5 percent running speed advantage for athletes with the higher belly buttons (i.e., black athletes) and a 1.5 percent swimming speed advantage for athletes with a lower belly button (i.e., white athletes).

As Jones pointed out, it would be blind and silly to ignore the importance of access to equipment and coaching. But this is a book about genetics and athleticism, and it would be just as blind to ignore the conspicuously thorough dominance of people with particular geographic ancestry in certain sports that are globally contested and have few barriers to entry. Namely, of course, that the athletes who are the fleetest of foot, in both short and long distances, are black.

9

We Are All Black (Sort Of)
Race and Genetic Diversity

You could carry a bag of blood onto an airplane in 1986. So the handoff that would help alter scientists' understanding of race and human ancestry took place at John F. Kennedy International Airport, in a rugged nook of Queens, New York.

Two colleagues of Yale geneticist Kenneth Kidd were traveling back from Africa with connecting flights at JFK, so he met them to collect the blood samples taken from the Biaka, a people from the Central African Republic, and the Mbuti, a people from the Democratic Republic of the Congo.

Growing up in Taft, California, the son of a gas station manager, Kidd had been fascinated by genetics since he was a twelve-year-old puttering around in the garden, marveling at what happened when he cross-bred different color irises. As an adult, he graduated to studying human DNA. Even before that handoff at JFK, Kidd had an inkling of what he would find.

At a scientific symposium in Italy in 1971 dedicated to the one hundredth anniversary of Darwin's *Descent of Man*, Kidd had presented data showing that some African populations have more variations—different possible spellings of the same gene or area of the genome—in their DNA than populations from East Asia or Europe. At the time, many scientists contended that Africans, East Asians, and Europeans had all reached the *Homo sapiens* stage independently; that *Homo erectus*—the precursor to modern man—had evolved separately on each continent to become the distinct ethnic variations that we see today.

Over the next two decades, Kidd filled his lab with the DNA of native populations spanning the globe. The Masai of northern Tanzania, the Druze of Israel, the Khanty of Siberia, the Cheyenne Native Americans of Oklahoma, Danes, Finns, Japanese, Koreans—all in translucent plastic containers color-coded by continent. Kidd collected some of the samples himself. Others, like DNA from the Hausa people of Nigeria, came from a Nigerian physician who was trying to figure out why women of certain ethnicities in southwest Nigeria give birth to twins at a higher frequency than women anywhere else on the planet.

Part of Kidd's goal was to categorize genetic variation around the world by looking at corresponding stretches of DNA in many different populations and examining how they differ. Every time he zoomed in on a portion of the double helix, a particular pattern held: more variation in the populations from Africa. For any piece of text in the DNA recipe book, there were almost always more possible spellings and phrasings in African populations than anywhere else in the world. In many areas of the genome, there was more genetic variation among Africans from a single native population than among people from different continents outside of Africa. On one particular stretch of DNA, Kidd observed more variation in one population of African Pygmies than in the entire rest of the world combined.

With geneticist Sarah Tishkoff, Kidd drew a family tree to represent everyone on earth. While African populations fanned out to form the bulk of the tree, all the European populations were clustered on tiny branches on the fringes. "From that genetic point of view," Kidd says, "I like to say that all Europeans look alike." This is because nearly the entirety of human genetic information was contained in Africa not so very long ago.

Kidd's work, along with that of other geneticists, archaeologists, and paleontologists, supports the "recent African origin" model—that essentially every modern human outside of Africa can trace his or her ancestry to a single population that resided in sub-Saharan East Africa as recently as ninety thousand years ago. According to estimates made from mitochondrial DNA—and the rate at which changes to it occur—the intrepid band of our ancestors who ventured out from Africa en route to populating the rest of the world might have consisted of just a few hundred people.

Humans split from our common ancestor with chimpanzees five or so million years ago. So relative to that time span, people have been outside of Africa for less—*much less*—than the equivalent of a two-minute drill in a football game. Because that band of our ancestors left not so very long ago in evolutionary terms, and took only a tiny fraction of the population along, they left behind most of humanity's genetic diversity. For millions of years, DNA changes had accumulated—both randomly and by natural selection—in the genomes of our ancestors inside Africa. But with only ninety thousand years for unique changes to occur outside of Africa, there simply hasn't been as much action in many stretches of the genome. People outside of Africa are descendants of genetic subsets of a group that was itself just a subset in Africa in the recent past.* Each time modern humans expanded to a new region of the globe, it appears that the pioneering emigrants were small in number and carried just a fraction of the genetic variation of their home en route to founding new populations. Data from around the world shows that the genetic diversity of native populations generally decreases the farther the population is along the human migratory path from East Africa, with populations native to the Americas tending to have the least genetic diversity.

---

*An important exception is that scientists have recently found that humans who ventured out from Africa must have interbred with Neanderthals, as the DNA of modern people in North Africa and outside of Africa—but not in sub-Saharan Africa—contains a small amount of Neanderthal DNA. While the picture of humanity outside of Africa as a subset of immigrants from a single African population is a general model, the more geneticists sample, the more complex is the story of genetic mixing that occurred both before and after humans traveled out from Africa.

This has momentous implications for classifying people according to their skin color. In some cases, the fact of an individual's black skin might indicate very little specific knowledge about his genome other than that he has genes that code for the dark skin that protects against equatorial sunlight. One African man's genome potentially contains more differences from his black African neighbor's than does Jeremy Lin's genome from Lionel Messi's.

There might also be implications for sports. Kidd suggests that for any skill that has a genetic component, theoretically, both the most and least athletically gifted individuals in the world might be African or of recent African descent, like African Americans or Afro-Caribbeans. Both the fastest *and* slowest person might be African. Both the highest *and* lowest jumper might be African. In athletic competition, of course, we seek to identify only the fastest runners and highest jumpers. "One can certainly find individual genes where there's more variation outside of Africa," Kidd says, "but the general picture is that there's more variation in Africa. . . . So you would expect out at the extremes there will be a greater proportion of people."

That said, there are clearly also *average* differences between populations, which is why Kidd does not recommend scouting for the next Olympic sprinter or NBA All-Star amid the staggering genetic diversity of African Pygmies. "There are certain anatomical features of the Pygmies that would intervene," Kidd says, referring to their extremely short stature. "But you might find the best basketball players in some of those populations in Africa where height and coordination are on average very high, and where you have a lot of other genetic variation within that group."

Kidd is suggesting that certain Africans, or people of recent African ancestry, *do* have a genetic advantage in sports performance at the upper end of athleticism. But because he is not professing an *average* genetic advantage, Kidd's supposition is intellectually palatable and has been touted as such both by scientists and in the press.

In the New Haven, Connecticut, lab that Kidd shares with his wife, Yale geneticist Judith Kidd, are the stainless steel refrigerators and garbage-bin-size liquid nitrogen containers that preserve the world's DNA, all neatly color-coded. The Yoruba people from Nigeria are there in their translucent yellow plastic box, the Han Chinese in a green box, and, in a purple box, Ashkenazi Jews. If Kidd had my DNA, it would be in the purple box.

In 2010, I had a portion of my genome analyzed by a private company that accurately traced my recent ancestry to eastern Europe and informed me that I carry a mutation on one copy of my HEXA gene. If I procreate with a woman who also carries the same mutation on one of her HEXA genes, each of our children would have a one-in-four chance of receiving two mutant versions of the HEXA gene and having Tay-Sachs disease, a nervous system disorder that results in death by age four. The HEXA mutation is uncommon throughout most of the world, but around one in every thirty Jews with Polish or Russian ancestors (like me) are carriers. The HEXA mutation is one among a batch of DNA signatures that make the people in Kidd's purple plastic box identifiable by their genes. Every one of the colorful boxes contains the DNA of populations with their own distinct genetic profiles.

"This is a genetic locus [a location on the genome] that affects how well you degrade Tylenol," Kidd says, through his handlebar mustache, as he clicks on a desktop file to open a study he coauthored. "There are certain mutations on this gene [CYP2E1] that cause acetaminophen poisoning of an individual." A rainbow-colored diagram appears on Kidd's monitor.

In this study, as in numerous others he has conducted, Kidd documented how common particular DNA spellings of sections of a gene are in fifty native populations from around the world. As expected, all sixteen of the spelling variations of CYP2E1 that Kidd examined—each represented by a different color—can be found in people in Africa, as can a number of other DNA spelling combinations that are found no-where else in the world. As the populations get farther from East Africa, through southwest Asia, Europe, northeast Siberia, Pacific Islands, East Asia, and the Americas, more and more of the colors drop out.

"You see, in Africa you have the lavender, the magenta, the yellow, the black, whatever," Kidd explains. "But when you get to Europe al-most everybody is going to have at least one copy of the green one." Among the Nasioi people, who are confined to the island of Bougain-ville in the Pacific Ocean near Papua New Guinea, every single mem-ber has the "green" DNA sequence in the CYP2E1 gene. "There are also Africans who have two copies of the green, so at that particular location [on the genome], one in a hundred Africans will be more similar to a European than another African," Kidd says. "But overall they're going to be very different from a European." Not only because they have unique spellings in their genetic code, but also because the frequency of gene variations is different in different populations. By looking at just one segment of one single gene, Kidd can start the process of homing in on a person's geographic and ethnic ancestry.

As ancestral humans spread across the world and became sepa-rated by all manner of obstacles—mountains, deserts, oceans, social affiliations, and later national boundaries—populations developed their own DNA signatures. For nearly our entire history, people lived, married, and procreated predominantly where they were born. As pioneers set up civilizations in new locales, gene variants became more or less common in populations both by random chance, or "ge-netic drift," as well as by natural selection when a version of a gene helped humans survive or reproduce in a new environment.

The gene variant that allows some adults to digest lactose, the sugar in milk, is one example. The general rule for mammals is that the lactase enzyme is shut down after the weaning period, and milk can no longer be fully digested. That held true for essentially all hu-mans just nine thousand years ago, before the domestication of cattle. Once humans kept dairy cows, though, any adult who could digest lactose was at a reproductive advantage, so gene variants for lactose tolerance spread like brushfire through societies that relied on dairy farming to thrive during winter, like those in northern Europe. Almost all present-day Danes and Swedes can digest lactose, but in populations in East Asia and West Africa, where cattle domestication is more recent or nonexistent, adult lactose intolerance is still the norm. Comedian Chris Rock famously joked that lactose intolerance is a luxury of wealthy societies: "You think anybody in Rwanda's got a f——g lactose intolerance?!" Rock asked in one of his routines. In fact, most people in Rwanda are lactose intolerant.

In an example particularly relevant to sports, about 10 percent of people with European ancestry have two copies of a gene variant that allows them to dope with impunity. The most common sports urine test that probes for illicit testosterone doping analyzes the ratio of testosterone to another hormone called epitestosterone—the "T/E ratio." A normal ratio is one-to-one. Injecting synthetic testosterone upsets the ratio by pushing the T higher than the E, and drug testers consider a ratio above four-to-one to signify possible cheating. But carriers of two copies of a particular version of the UGT2B17 gene pass the test no matter what. The gene is involved in testosterone excretion and one version of it causes the T/E ratio to remain normal no matter how much testosterone one injects. So 10 percent of European athletes can cheat and still have no chance of failing the most common drug test. And the get-out-of-drug-testing-free gene is more rule than exception in other parts of the world, like East Asia. Two thirds of Koreans have the genes that confer immunity to T/E ratio testing.

Despite our differences, because all humans have common ancestry that is not so distant in the past, we are exceedingly similar, more similar across the entire genome than chimpanzees are to one another. At the DNA level, of the three billion letters in the recipe book, humans are generally about 99 to 99.5 percent the same. In a sense, you probably knew that intuitively. If you had to build two human beings from scratch, no matter where in the world they were from, most of the instructions would be identical: two eyes, ten fingers and toes, a liver and two kidneys, all the same bones and brain chemicals. For that matter, just about every page would be the same for a human and a chimp, as we are 95 percent similar to chimps at the DNA level. But it is a mistake to take all that to mean that the differences are unimportant.

At least 15 million letters of the DNA code differ on average between individuals, and the actual length of people's genomic recipe book can differ by millions of letters as well. It is plenty enough difference to cause all the variation we see in the world. In 2007, as genome sequencing became faster and cheaper, *Science,* one of the two most prestigious scientific journals in the world, named as its breakthrough of the year the revelation of "how truly different we are from one another" at the genetic level. As genome sequencing has become cheaper still, that point has only been amplified. Wherever humans have set up civilizations, they have rapidly differentiated themselves.

Though natives have inhabited Iceland for just a single millennium, the company deCODE Genetics showed that it could identify which of eleven regions of Iceland a resident's grandparents hailed from using just forty areas along the genome. In 2008, scientists looking at much larger swaths of DNA pinpointed the geographic ancestry of nearly all of a sample of three thousand Europeans to within a few hundred miles. And, to a degree, DNA can identify the construct we call "race" as well.

A 2002 study published by a team of researchers (including Kidd) in *Science* directed a computer to peruse 377 spots on the genomes of 1,056 people from around the world and then automatically separate the people into groups based on genetic differences. The groups that the computer delineated corresponded with the world's major geographic regions: Africa, Europe, Asia, Oceania, and the Americas. A subsequent Stanford-led study asked 3,636 Americans to self-identify as either white, African American, East Asian, or Hispanic, and found that the self-identification matched a blind DNA identification in 3,631 cases. "This shows that people's self-identified race/ethnicity is a nearly perfect indicator of their genetic background," geneticist

Neil Risch said in a press release issued by the Stanford University School of Medicine.[*]

Skin color, which is primarily determined by latitude, can be an imprecise marker of geographical ancestry, as there are spectrums of skin color on each continent. But geography and ethnic affiliation have most certainly left a trail of genetic crumbs.

In some areas of medicine, like pharmacogenetics—the study of how and why people with different genes respond differently to the same drugs—skin color is already being used as a proxy, albeit often a crude one, for underlying genetic information, and medical researchers now recognize the importance of testing the efficacy of drugs separately on different ethnic groups.

In 2004, Kidd and Tishkoff wrote that the main genetic and geographic clusters of people do "correlate with the common concept of 'races,'" but added that if every population on earth were included, the genetic differences would look more like a continuous spectrum as opposed to a collection of discrete groups.

In 2009, Tishkoff and an international team published a landmark study that characterized the genetic backgrounds of African Americans. They found that adults who identify as African American are highly genetically diverse on the whole, with ancestry ranging from 1 percent to 99 percent West African. African Americans are particularly diverse in the amount of European ancestry they have in their DNA. But almost all African Americans were found to have African X chromosomes, consistent with the idea that the mothers of African Americans have historically been of very recent African ancestry, while fathers were sometimes African and sometimes European. The African Americans studied were from Baltimore, Chicago, Pittsburgh, and North Carolina, and the African components of their genetic ancestry showed "little genetic differentiation," according to Tishkoff, and were similar

---

[*]It should be noted, however, that African Americans are predominantly from a specific swath of Africa.

to one another and often to the genetic profiles of West African people like the Igbo and Yoruba of Nigeria; not a surprise, as the Igbo and Yoruba show up frequently in records of the slave trade as Africans who were wrenched from their homes and taken to the Caribbean and the United States.*

One's ancestry can be traced through one's genes, but to go further down the path of Kidd's thought experiment about African athletes, we must know not only that the genotypes of African people are the most diverse, but whether their *phenotypes* are also the most diverse. A phenotype is the physical manifestation of underlying genes. Geneticists still have scant clue what most of the billions of bases (each "letter" is one base) in our DNA do. Some may do little or nothing at all. Kidd's suggestion is that because the greatest diversity of genotypes is contained in African populations, the greatest diversity of athletic phenotypes—both the slowest and the fastest runners—might also be there. So far, though, there is no easy, blanket conclusion to Kidd's thought experiment.

In 2005, the U.S. government's National Human Genome Research Institute weighed in on the issue of race and genetics and the question of whether most of the physical variation in the world occurs among individual people within ethnic groups, or among entire ethnic populations themselves. The institute directly addressed the question of whether the high degree of genetic diversity in African populations also means that most of the physical diversity in the world is contained in those populations. The answer: it depends on the specific physical trait you're looking at.

About 90 percent of the variation in the shape of human skulls occurs within every major ethnic group—only 10 percent separates ethnicities—with Africans indeed showing the greatest variation. But the exact opposite is true for skin color: only 10 percent of the variation

---

*A separate genetic study of African Americans found that those in South Carolina tended to hail from the "Grain Coast"—Senegal to Sierra Leone—probably because rice plantation owners in South Carolina wanted slaves skilled in a particular kind of agricultural work that was desirable in South Carolina.

occurs within ethnic groups, and 90 percent of the difference is between groups. Thus, in order to discuss whether Africans or African
Americans have specific genes that are advantageous in certain sports,
scientists should first identify specific genes and innate biological traits
that are important for sports performance, and then examine whether
they occur more frequently in some populations than in others.

They have begun to do just that.

Kathryn North had the letter to *Nature Genetics* all ready to go, and her
report would be a breakthrough.

A few years earlier, in the summer of 1993, North had left Australia
to train as a pediatric neurologist and geneticist at Boston Children's
Hospital, where she worked in a lab that had discovered the genetic
mutation that causes Duchenne muscular dystrophy, a devastatingly
virulent muscle-wasting disease. When North examined the muscle
fibers of muscular dystrophy patients, she saw that they had a normal
ration of fast-twitch muscle fibers, but that about one in five patients
was missing a particular structural protein called alpha-actinin-3 that
should have been in those explosive muscle fibers.

Her letter to *Nature Genetics* would document the case of two Sri
Lankan brothers North examined in her lab in Sydney in 1998 and
who had congenital muscular dystrophy. The brothers' parents, who
did not have the disease, were cousins, so the case appeared to be one
of recessive genetic inheritance. Neither of the boys had any alpha-
actinin-3, so North and colleagues sequenced in each boy the gene
that codes for it, the ACTN3 gene. Sure enough, each boy had a "stop
codon," a genetic stop sign, at the same spot on both copies of the
ACTN3 gene. The stop sign—just one single letter switch in the
DNA—prevented the alpha-actinin-3 protein from being produced in
muscles. North and her team, it appeared, had discovered a new gene
mutation that caused muscular dystrophy. "I started drafting a letter
to *Nature Genetics,* and I was literally drafting a paper to report a new

disease gene," she says. "But if you're a good geneticist, you bring in the whole family."

So North invited the parents and their other two, healthy children and probed their ACTN3 genes too. The version of the gene that the ill brothers had that stopped alpha-actinin-3 production is known as the X variant, and North expected the parents each to have one X variant, which they had passed to their sons, and one R variant, which functioned normally and facilitated production of the protein. To her surprise, both parents and the two healthy siblings also each had two X variants of the ACTN3 gene. Nobody in the family had any alpha-actinin-3 in their muscles whatsoever, yet only the two brothers had muscular dystrophy. North had not found a new muscular dystrophy gene after all. "That was a Friday when we found out," she says, "and it was really, really depressing."

That Sunday, she went to a movie and afterward took a walk to ponder the previous week. Never, not in the lab nor in the scientific literature, had she found an example of a healthy person with genes that left them entirely devoid of a structural protein. Structural proteins are critical. They make fingernails, hair, skin, tendons, and muscle. Humans tend to be diseased or to die when the genes that code for them are not functioning. "So I started reading the evolution literature," North says, "and I thought, well, maybe alpha-actinin-3 is redundant. Maybe we don't need it and it's on its way out."

North cold-called Simon Easteal, an Australian researcher with a specialty in molecular evolution. Together they yanked from storage two hundred samples of muscle with all manner of disease, from muscles that did not contract properly to others that had malfunctioning nerves. Just as she had seen with muscular dystrophy patients in Boston, about one in five of the diseased muscles had two copies of the X version of the ACTN3 gene, and thus no alpha-actinin-3. But about one in five samples of normal, healthy muscle had two X variants as well, so the gene could not be the cause of disease. Perhaps, then, alpha-actinin-3 had some other purpose in muscle. "That's when we started to pull in different groups of people," North says. "And that's when we found this different ethnic distribution of the gene."

North saw that one quarter of people of East Asian descent had two copies of the X variant of ACTN3 and about 18 percent of white Australians had two X variants. But when she tested Zulu people from South Africa, less than 1 percent had two X variants. Nearly all had at least one copy of the R variant, which codes for alpha-actinin-3 in fast-twitch muscle. And that was true of every African population. With respect to this particular gene variant, Africans or people of recent African ancestry happen to be extraordinarily uniform.

North was convinced that alpha-actinin-3 was not a meaningless protein, even though its absence did not lead to disease. Like the myostatin protein—of Superbaby fame—alpha-actinin-3 was highly conserved in evolutionary terms. It is in the explosive muscle fibers of chickens, mice, fruit flies, and baboons, among other animals, including our closest primate relatives, chimps. The absence of alpha-actinin-3, then, is a very recent and very human trait. North and colleagues estimated that the X variant spread through humans within the last thirty thousand years, and only outside of Africa. The gene, it appears, had been favored by natural selection only in non-African environments for some reason. Fast-twitch fibers must need it for something, North thought.

So she and her colleagues collected DNA from subjects with ample fast-twitch fibers: elite sprinters. They partnered with the Australian Institute of Sport to do ACTN3 testing on international-level athletes. While 18 percent of Australians had two X copies of the gene, almost none of Australia's competitive sprinters did. Nearly every sprinter produced alpha-actinin-3 in their fast-twitch fibers. "I waited for years to publish that study," North says. "The result came out the first time we did the analysis, and then we repeated it again and again internally." And it held. Not only did sprinters in general tend not to have two X copies of ACTN3, but the better they were, the less likely it was they were XX. In one sample, just 5 out of 107 Australian sprinters were XX, and zero of the 32 sprinters who had gone to the Olympics were XX.

After that work was published, sports scientists around the world hustled to test their local sprinters, and the association showed up everywhere. With almost no exceptions, sprinters from Jamaica and Nigeria all had alpha-actinin-3 in their fast-twitch muscles, but so did distance runners from Kenya—no surprise given that nearly all of the control subjects from African populations did as well. Scientists in Finland and Greece took DNA from their Olympic sprinters, and, again, not a single one was XX. In Japan, a few sprinters were XX, but none who had run faster than 10.4 seconds for 100 meters.

ACTN3, North concluded, is a gene for speed. Why that may be so is not exactly clear. Alpha-actinin-3 may have a structural impact on how explosively a muscle fiber can contract, or it may influence the configuration of the muscular system. Mice as well as—in several studies—Japanese and American women who are deficient in alpha-actinin-3 have smaller fast-twitch muscles and less muscle mass over all. When North bred mice to have no alpha-actinin-3, compared with normal mice they had far less active glycogen phosphorylase, the enzyme that mobilizes sugar for explosive actions, like sprinting. The fast-twitch muscle fibers in those mice also took on some of the properties of slow-twitch, endurance fibers.

Given the approximate timing of when the X version of ACTN3 appears to have spread through humans—fifteen to thirty thousand years ago—North has toyed with the idea that the variant may have proliferated during the last ice age. Absence of alpha-actinin-3 may make fast-twitch muscle fibers more metabolically efficient, like their slow-twitch neighbors, a boon, perhaps, in frigid, food-scarce northern latitudes outside of Africa. Two anthropologists have suggested that the X version may have spread when humans outside of Africa transitioned from the hunter-gatherer lifestyle to an agricultural one, where they would have had less need to sprint in war or hunting but more need to be metabolically efficient and to work at a steady rate for long hours.

But North is cautious. Though we share the vast majority of our DNA sequence with mice, genetically manipulated rodents are not ideal models of human genetic variation. "We don't know the whole story," North says. "Right now, it looks like ACTN3 is one gene that contributes a little to sprinting, and there may be hundreds, and of course there are other factors like diet, environment, and opportunity."

Private gene-testing companies have been less circumspect. As soon as the ACTN3-and-Olympians study appeared, companies rushed into the sparsely regulated direct-to-consumer genetic testing market. Genetic Technologies, of Fitzroy, Australia, led the way. For $92.40, the company would tell a customer what versions of the ACTN3 gene they carry. (I have two R copies.) In 2005, the Manly Sea Eagles of Australia's National Rugby League became the first team to admit publicly that it was testing players for ACTN3 and tailoring training programs accordingly, giving more explosive weight lifting and less cardio to the guys with sprinter variants.

Atlas Sports Genetics, of Boulder, Colorado, has made headlines for selling parents an ACTN3 test for their children. According to Kevin Reilly, president of Atlas, the test is particularly useful for "those younger athletes who don't have the motor skills yet." By "younger," Reilly means that just because baby Kobe doesn't know how to walk yet, that should not mean his DNA can't begin charting his athletic career. If Kobe has no R versions of the gene, his parents can start nudging their little bundle of DNA toward endurance sports. The genetic-testing-in-diapers market barely materialized for Atlas, but the company did manage a preteen customer base. Says Reilly, "We have had some impact on athletes in the eight-to-ten age group," in terms of influencing their sport choices.

Unfortunately for those eight-to-ten-year-olds, though, consumer genetic testing for athleticism is nearly worthless.* Scientists have in-

---

*In fairness, several studies that tested elite endurance and sprint/power athletes for a set of genes associated with either endurance or power found, in general, that a gene panel could distinguish the endurance athletes from the sprint athletes. (But any coach worth her salt could do this with greater accuracy.) A 2009 study of Spanish sprinters and jumpers tested athletes for six gene variants associated with explosiveness. Five out of fifty-three of the athletes had all six of the "power" versions of the genes, whereas that should occur in only one in every

creasingly realized that the inherited component of complex traits, like athleticism, is most often the result of dozens or even hundreds or thousands of interacting genes, not to mention environmental factors. If you are XX for the ACTN3 gene, "you probably won't be in the Olympic 100-meters," North says. But you already knew that, without a genetic test. Though the ACTN3 gene does appear to influence sprinting ability, making a sports decision based on it is like deciding what a puzzle depicts when you've only seen one of the pieces. You need that piece to complete the puzzle, but you certainly can't see a meaningful picture without more pieces.

As Carl Foster, director of the Human Performance Laboratory at the University of Wisconsin–La Crosse and coauthor of several ACTN3 studies, puts it: "If you want to know if your kid is going to be fast, the best genetic test right now is a stopwatch. Take him to the playground and have him race the other kids." Foster's point is that, despite the avant-garde allure of genetic testing, gauging speed indirectly is foolish and inaccurate compared with testing it directly—like measuring a man's height by dropping a ball from a roof and using the time it takes to hit him in the head to determine how tall he is. Why not just use a tape measure?

All that ACTN3 can tell us, it seems, is who will *not* be competing in the 100-meter final in Rio de Janeiro in 2016. And it is not even doing a very specific job of that, given that it is only ruling out about one billion of the seven billion people on earth.

Still, if only that one gene is taken into account, it is also telling us that there are almost no black people anywhere in the world who are ruled out.

---

five hundred normal Spanish men. Interesting for research purposes, but still not very useful for predicting whether a child will become a sprinter or jumper or marathon runner.

10

The Warrior-Slave Theory of Jamaican Sprinting

Welcome home again!" the black scientist says to the white scientist, a Cheshire Cat smile curling around his face.

The black scientist is Errol Morrison, the most renowned medical researcher in Jamaica. "Morrison Syndrome" is a form of diabetes that he linked to indigenous bush teas that some Jamaicans consume in copious quantities. Morrison is so esteemed on the island that once when he was receiving an award for his work, the doctor introducing him joked to the audience that when she traveled abroad people who learned she was from Jamaica would greet her with "Bob Marley!"—unless it was a diabetes conference, in which case they say, "Errol Morrison!"

Morrison is also the president of the twelve-thousand-student University of Technology in Kingston, known locally as UTech. And right now, in late March 2011, he's joking with the white scientist, Yannis Pitsiladis, a biologist and obesity expert from the University of Glasgow who visits the island regularly and was recently made an adjunct distinguished professor in UTech's nascent sports science program.

Now the men's right hands are clasped, and each has his left around the other man's back. There is a glistening affection between them. They will relax over dinner tonight in Morrison's airy home, high on a hill, with the Kingston lights just pinpricks below.

But Pitsiladis is in town to work. For a decade now, he has been traveling here with cotton swabs and plastic containers asking for bits of cheek and gobs of drool from the planet's fastest men and women. There is no place else on earth where he's liable, over lunch, to bump into a half-dozen men and women who ran in the Olympic 100-meters. When he does, he will be sure to collect their DNA. (Once, during a chance encounter with a world-class runner at a social function, Pitsiladis hastily sterilized a wineglass for saliva collection.) UTech itself, with its humble, 300-meter grass track, is a hotbed for speed. Sprinters and jumpers who trained at UTech won more medals in track and field (eight) at the Beijing Olympics than dozens of entire countries won in the entire Games.

Over dinner, Morrison and Pitsiladis will talk about their shared scientific goal: untangling the factors, genetic and environmental, that have made a tiny island of three million into the world's sprint factory. They have put their formidable brains together, and they have published papers together. They have also published separately on the topic in the scientific literature.

And the conclusions of those papers, on the issue of nature-or-nurture, could hardly be more opposite.

In his memo pad for work-related expenses, Pitsiladis has a budget line for paying a witch doctor in Jamaica in his quest to get approval to collect DNA from the man's community. Needless to say, there are few researchers like him in the world.

Pitsiladis's ancestors left Greece after World War II in search of work, moving first to Australia, and then South Africa. From 1969, when he was two, Pitsiladis lived in the land of apartheid. In 1980, his family returned to Greece, to the island of Lesvos, where he obsessed over training for a career as a professional volleyball player. The future biologist cut school to practice, but when he topped out at 5'10", Pitsiladis surrendered his volleyball dream. Both his previous lives, in South Africa and Greece, can be found embedded in the work he does now: looking for genes that make the planet's best athletes, and asking whether one ethnicity has cornered the market on that precious DNA. For a decade, that has meant traveling to Ethiopia, Kenya, and Jamaica, to the training grounds of some of the most endurant and most explosive athletes on earth.

The work has been arduous. Time and again, Pitsiladis has been denied funding to examine the genes of athletes, as research funding for human genetics is generally earmarked for the study of human ancestry or health and disease. So Pitsiladis sustains his academic position at the University of Glasgow by studying the genetics of childhood obesity, a line of inquiry that attracts hefty grants. Pitsiladis's dean at Glasgow has made a point of telling him to ditch the athlete work and focus on his obesity research. But Pitsiladis is maniacal about his research passion, and obesity genetics is not it.

"I just published a paper on a fat gene," he says, "but [the gene] has a very small effect, and that can be overcome with physical activity. And we'll find many more genes, and already I can tell you what the answer will be." He holds up his thumb and forefinger, an inch apart. He's indicating that although scientists will find dozens, or hundreds, or thousands of DNA variations that contribute to a predisposition for being overweight, they will all amount to only a small fraction of the explanation for the industrialized world's obesity epidemic.

It is as if Pitsiladis peels off a dour mask when he switches from discussing obesity genetics to his other work: peering into the genes of the greatest athletes in the world. He occasionally dons a gold and green Ethiopia track-and-field shirt, a gift from an Ethiopian gold medalist, and strands of salt-and-pepper hair bounce off his temples when he gets excited. His eyelids peel back, and his delicate accent, an amalgam of the countries where he has lived, leaps to mezzo-soprano. "My brain never switches off this topic," he says. "It never stops. Never. I once worked for a year to get one DNA sample! Who else is going to do that?" The answer, in sports science: no one, because there is scant funding for it.

And so Pitsiladis's sports research must proceed via the bubble gum and duct tape school of science. Since he started visiting Jamaica in 2005, Pitsiladis has paid for much of the work from his own pocket (he remortgaged his home, twice); by collaborating with media (he sold footage from Jamaica to the BBC for a documentary); by partnering with foreign scientists (the Japanese government has carved out a bit of funding for sports genetics); and with a little help from his friends—a 2008 trip to Jamaica was funded by the owner of Pitsiladis's local Indian restaurant in Glasgow, on the condition that the restaurant owner's son be allowed to tag along.

This is science at its most wondrously bold and shoestring. And still, for Pitsiladis, getting the funding can be as harrowing as not getting it. He is deathly afraid of flying. His assistant can expect a call prior to every visit to Africa or Jamaica, the man on the other end pleading for the trip to be canceled. But with the help of some vintage red, he always boards.

Not all of his trips to Jamaica have revolved around DNA collection. On the first few visits, Pitsiladis was more of an anthropologist, asking the Jamaican people themselves for their own theories on the secrets of the sprint factory. Answers spanned from the yams they eat to rural children's habit of chasing animals, to the people's history of sprinting away from European slave masters. The latter idea may sound silly, but it has origins as deep as the caverns of northwest Jamaica, the locale from which it springs.

Early in his Jamaican ventures, Pitsiladis learned that not only does the island produce an extravagant number of the world's top sprinters—the national 100-meter record holders for Canada and Great Britain are Jamaican expats, and top American sprinters often have Jamaican roots—but many hail from in and around the tiny parish of Trelawny, in Jamaica's northwest quadrant. The 2008 Beijing Olympics were the crowning achievement of sixty years of Jamaican sprint success. And the '08 winners of both the Olympic men's 100- and 200- and the women's 200-meter dash—Usain Bolt and Veronica

Campbell-Brown, the premier sprinters of a generation—hail from Trelawny. In the eighteenth century, it became home to a small band of unlikely warriors who descended the sheer limestone cliffs from the thickly layered rain forest of Jamaica's Cockpit Country into the valleys below to terrorize the most refined soldiers of the world's most feared military.

It is these Jamaican warriors, Pitsiladis was told, who spawned today's captains of track and field.

On April 3, 2011, one week after his gourmet dinner with Morrison, Pitsiladis is sitting in a chipped plastic chair in a dimly lit concrete room in the rain-forested region of Jamaica that most of the island's natives have never seen. And he is fighting for his science.

Across a wooden desk that was dragged into place for this meeting is Colonel Ferron Williams, the leader of Accompong Town. Williams is wearing a golden brown short-sleeve button-down, and his perfectly shaved head tilts quizzically as he listens. To his left is Norma Rowe-Edwards, his deputy and the town nurse.

When Pitsiladis visited three years ago to gather DNA from Accompong residents, Rowe-Edwards voiced concern about his collection method because it necessitated rubbing cotton swabs inside mouths. Within days, the gossip around Accompong was that Pitsiladis's cheek swabs were spreading AIDS.

To the colonel's right is a local man whom Pitsiladis hired in 2008 to help with the swabbing. The man promised to collect DNA from two hundred Accompong natives. But when Pitsiladis returned to Glasgow to analyze the data, the sequence of Gs, Ts, As, and Cs was the same in all two hundred samples. The man claimed that area residents must just be very closely related. But the sequence was not *close*, it was *identical*. The man had swabbed himself two hundred times.

Despite these previous travails, embodied by the people now at the table, Pitsiladis is prevailing in today's discussion. The DNA collection kits no longer require a swab, just drool in a plastic disc, so the nurse's concerns about invasive testing are alleviated. And the colonel would like to draw attention, and visitors, up the lone, spiraling mountain road that leads to this tiny farming community with its low-lying, pastel-colored concrete structures placed haphazardly beside canted tin shanties. So he's glad to watch over the scientific work so that it can proceed without obstacle. By the end of the meeting, the colonel has reached across the table to grasp Pitsiladis's hand. He has given his permission for more sampling.

This wedge of Jamaica is of paramount significance to Pitsiladis. The oral history of northwest Jamaica tells that the fiercest slaves were brought here, first by the Spanish and eventually the British, because it is surrounded by cliffs and ocean and difficult to escape. The part of the story that drew Pitsiladis here begins in 1655, when the British navy came to Jamaica to wrest control of the island from the Spanish. Intrepid slaves took advantage of the chaos to flee into the Cockpit Country, the mountainous highlands of northwest Jamaica. The escaped slaves founded their own communities and became known as Maroons, from the Spanish word *cimarrón*, which describes domesticated horses that flee into the wild.

The geography of Jamaica's Cockpit Country is entirely unique on the island, and rare in the world. Known as karst topography, the remote and wet forest blankets limestone that has been cut away by millions of years of rain, leaving star-shaped valleys—called cockpits—walled in by sheer, vertiginous cliffs. Unlike most valleys formed by water, these have no rivers. The water works its way through the porous limestone and disappears into a lattice of underground caverns. For Maroons who mastered the terrain, and knew the layout of the limestone sinkholes, the Cockpit Country provided an impregnable defense against British troops.

After taking over from the Spanish, the British furiously ramped up slave importation, bringing Africans by the thousands from locations that correspond to modern-day Ghana and Nigeria. Many came from ethnic groups expert in warfare—like the Coromantee of Ghana—sometimes sold into slavery by rival peoples who captured them. Contemporary letters from British officials show deep respect for the Coromantee, whom one British governor in Jamaica called "born Heroes . . . implacably revengeful when ill-treated," and "dangerous inmates of a West Indian plantation." Another Brit, writing in the eighteenth century, said that these "Gold Coast Negroes" were distinguished by "firmness both of body and mind; a ferociousness of disposition . . . an elevation of the soul which prompts them to enterprizes of difficulty and danger."

In the 1670s, as slaves were increasingly brought to Jamaica, and increasingly fled to join the burgeoning communities in the mountains, the Maroons burned sugarcane fields, painting the night sky with the color of their intentions. "No flame is more alarming" than a cane fire, wrote William Beckford, an Englishman living in Jamaica. "The fury and velocity with which it burns and communicates cannot possibly be described." From those bold Coromantee came the military genius known as Captain Cudjoe.

Cudjoe, along with Nanny, the female leader of Maroons on the east side of the island, created an elaborate spying system that employed Maroon soldiers and slaves on plantations to track the movements of British soldiers.* When the British ventured into the Cockpit Country to retrieve runaway slaves, Cudjoe's fighters ambushed them, not merely beating them back despite their superior numbers, but building an army with the weapons they seized. The battles were so lopsided that the soldiers of the vaunted British Empire, wrote one English planter, "dare not look [the Maroons] in the face . . . in equal numbers." That British dread is still embedded in the local names of Cockpit Country districts: *Don't Come Back* and *Land of Look Behind*.

The climactic battle occurred in 1738, just a short stroll from where

---

*Nanny is so revered in Jamaica that island legend has it she could snatch British bullets out of the air.

Pitsiladis met the colonel to discuss DNA collection. A band of Cudjoe's soldiers hid in a limestone cave, now called Peace Cave, and placed a loose rock on the path outside. The British soldiers clattered the rock as they passed, while the Maroons waited, tallying their number. Then one of the Maroons emerged and signaled to others in the surrounding hills by blowing into an *abeng*, a bellowing instrument carved from a painted-green cow horn. Maroon fighters flooded the valley from every direction and massacred the British soldiers. Legend has it that only one British soldier was spared, sent home with his ear in his hand, to tell his superiors what had occurred. Shortly after the slaughter the British signed a treaty with the Maroons, granting them their remote territory—Cudjoe was made chief commander of nearby Trelawny Town—and their freedom, a full century before official emancipation.

Today, the five hundred or so Maroons of Accompong Town comprise a sovereign nation inside of Jamaica. Just over the hill from where Pitsiladis and the colonel met are the childhood homes of Usain Bolt and Veronica Campbell-Brown.* Maroons in Accompong Town do not hesitate to claim them as members of their lineage.

"No one can argue that there was selection of the fittest slaves," Pitsiladis says. He has seen some of the historical records himself, interviewed experts on the island, and coauthored papers on the demographics of the Jamaican slave trade. "The guys selling the slaves were their neighbors," he says. "What happened was: I knew you were strong, and before you knew it I had a hood over your head, and I sold you. So, eventually, the strongest and fittest got on those ships." And the strongest and fittest of those supposedly ended up in the northwest quadrant of the island, as indomitable Maroons. "And that's the

*Perhaps the most infamous sprinter of all time, Canada's Ben Johnson, who won gold in the 100 at the 1988 Olympics only to be stripped days later after he tested positive for steroids, also hails from Trelawny.

area where the athletes are from in Jamaica," Pitsiladis says. "So it all makes a really convenient story."

The story: that strong people were taken from Africa; that the strongest of those survived the brutal voyage to Jamaica; that the strongest of those strong fed the Maroon society that cloistered itself in the most remote region of Jamaica, and that the Olympic sprinters of today come from that isolated, warrior genetic stock. (In a 2012 documentary, world-record-holding sprinter Michael Johnson sided with the theme of that theory: "It's impossible to think that being descended from slaves hasn't left an imprint through the generations. . . . Difficult as it was to hear, slavery has benefited descendants like me— I believe there is a superior athletic gene in us.")

Since 2005, Pitsiladis has collected DNA from Maroons as well as 125 of the best Jamaican sprinters of the past fifty years. (He is careful not to identify exactly the athletes he has taken genetic material from. When I visited Pitsiladis's lab in Glasgow, he hovered over a grad student who was using a pipette to transfer the DNA of "the likes of a Usain Bolt," Pitsiladis said, onto a plastic sample plate.)

His data, though it is preliminary, has not particularly supported the idea that the Maroon warrior society specifically spawned the Jamaican sprint society.

Maroons in Accompong Town repeatedly told me they could pick other Maroons from a crowd by the darkness of their skin. But, when pressed, most admitted that this was just a bit of folklore they repeat, and that they actually probably could not do so. Nor can Pitsiladis, from the standpoint of their DNA—though he has analyzed only a fraction— really tell the Maroons apart from other Jamaicans. "They look [genetically] like West Africans, and so do all the other Jamaicans," he says. "Look around, and try to tell me, what *is* a Jamaican?"

Pitsiladis is referring to the fact that the DNA of Jamaicans follows their national motto: "Out of Many, One People." Slaves came to Jamaica from a raft of countries in Africa, and from a bundle of ethnic groups within those countries. Genetic studies of Jamaican ancestry have found an array of West African lineages. One study of a section of Jamaicans' Y chromosomes—passed only from fathers to sons—found that they tended to be most similar to Africans from the Bight of Biafra, which includes coastal areas of Nigeria, Cameroon, Equatorial Guinea, and Gabon. An investigation of Jamaicans' mitochondrial DNA found more similarity with Africans from the Bight of Benin and the Gold Coast, which include areas in Ghana, Togo, Benin, and Nigeria. As with African Americans, all of the studies agree that the genetic matrilines of Jamaicans are essentially entirely West African, but from a number of countries.

In short, as expected from the island's slave importation history, the residents descend from western Africa, but from a variety of ethnic groups therein. (Captain Cudjoe, after all, famously united fighters from the Ashanti, Congolese, and Coromantee tribes.) Not to mention that genetic studies have found that some Jamaicans carry a bit of Native American DNA, presumably from mixing with the Taino people, native inhabitants of Jamaica who some historians previously thought went extinct from disease and persecution at the hands of Spanish colonizers before West African slaves arrived.

Colin Jackson, who held the 110-meter hurdles world record from 1993 to 2006, has Jamaican parents but was born and raised in Wales. He underwent genetic analysis in 2006 for the BBC ancestry program *Who Do You Think You Are?* To Jackson's surprise, his DNA revealed that he is 7 percent Taino. Historians now believe that a small number of Taino people must have survived the Spanish occupation by fleeing into the hills to join the Maroons. So the British Jackson may be yet another world champion sprinter with Maroon heritage. (In 2008, five years after his retirement, Jackson participated in another BBC program, *The Making of Me*, in which a laboratory at Ball State University took a sample of muscle tissue from his leg and determined—to Jackson's utter delight—that he had the highest proportion of type IIb, or "super fast twitch" muscle fibers, that the lab had ever seen.)

Clearly, there are intricacies yet to be discovered regarding the genetic heritage of Jamaicans as well as the island's premier sprinters. But, at the least, the work of Pitsiladis and others has shown that neither the Maroons nor Jamaicans overall constitute any sort of isolated, monolithic genetic unit. Rather, as we should expect from a mixed group of West Africans, Jamaicans are highly genetically diverse. (Though, also as expected, Jamaicans are decidedly *not* diverse when it comes to the ACTN3 "sprint gene." Nearly all Jamaicans have a copy of the right version for sprinting.)

If the sprint factory phenomenon came down to the natives of the sprint-happy Caribbean with the highest degree of genetic African-ness, then we would expect more top sprinters from Barbados, as the 250,000 inhabitants of that tiny island tend to have the least diluted West African ancestry in the Caribbean. (That said, Barbados actually *is* overrepresented given its population—an Olympic medalist in the 100-meters in 2000 and an Olympic finalist in 110-meter hurdles in 2012—though not to the degree of Jamaica. The tiny Bahamas, population 350,000, is also perennially one of the best sprint countries in the world. Bahamas beat the United States to win gold in the men's 4×400 relay at the 2012 Olympics. Trinidad and Tobago, population 1.3 million, is yet another of the Caribbean's global sprint powerhouses.)

When Pitsiladis compared two dozen gene variants that have been associated with sprint performance—albeit extremely tenuously in some cases—in Jamaican sprinters and control subjects, the results "went in the right direction," he says, "but it was not dramatic." That is, sprinters did tend to have more of the "right" versions than nonsprinters, but it was by no means always the case. One of Pitsiladis's grad students, who was used as a control subject, had more of the sprint variants than "the likes of a Usain Bolt." This does not mean that genes are unimportant for sprinting, but rather that scientists have located only a very small number of the relevant genes.

Pitsiladis continues to analyze the genes of top Jamaican sprinters, and as technology has made it easier to study vast swaths of the genome, a few gene variants have emerged in his work as differing between sprinters and controls and thus as potential influences on sprinting success, but the story is murky. And because there are too few Olympic-medal-caliber sprinters in the entire world to create large studies, it will likely remain murky. Sports scientists have a tortuous path ahead to uncover many of the physical qualities that lead to elite athletic performance, much less the genes that undergird them.

In his decade of travel to Jamaica, Pitsiladis's theories regarding the world's sprint factory have been influenced less by the data he has compiled with expensive DNA sequencers and chromatographs, and far more by the data he has gathered with two other important scientific instruments: his eyeballs.

Known simply as "Champs," Jamaica's national high school track-and-field championships has run continuously since 1910—when Jamaica was still a British colony and the headmasters of six boys' schools arranged the races—and it is the island's crowning entertainment event of the year.

Champs extends over four days with one hundred high schools in both the boys' and girls' competitions. The riotous final day is what you might get if all at once a thousand nightclubs were poured into a track meet.

Kingston's 35,000-seat National Stadium becomes standing room only, with enough dancing fans in the aisles to show off the "whine," a methodical, hip-rolling dance that will coax a blush from the uninitiated. In the evening, the stadium halls are redolent of jerk seasoning, and seating areas filled with devotees of a particular high school are covered in brightly colored banners the size of schooner sails. When the fans spot a "cracker"—local jargon for a hotly contested race—the noise of cheers and whoops and whistles and horns crescendos to deafening as athletes lean for the finish in lockstep. If an anchor leg in a sprint relay starts catching a competitor late in a race, the PA announcer will remind spectators not to show their excitement by jumping out of the stands onto the track. Olympic sprinters show up to cheer on their old schools or to bask in celebrity. At the 2011 Champs, a retinue of girls in sequined shirts and boys in open jackets and loose sneakers bulged around Asafa Powell as the former world record holder—in designer jeans, gold chains, and sunglasses at night—sauntered through the stands.

Youth track is all the rage in Jamaica. Prior to Usain Bolt, professional track meets in Kingston played to empty stands, outdrawn even by the national championship for five- and six-year-olds. Puma stores around Kingston stock gear emblazoned with the emblems of schools that boast hallowed Champs histories, like Calabar High, named after a port city in Nigeria that was a final point of departure for slaves. The fever pitch of youth track gives rise to enthusiasts who want to help their local school succeed at Champs. Enthusiasts like Charles Fuller.

Back in 1997, when he was an employee of the Alcan Jamaica aluminum company, Fuller was sick of watching the fastest local kids leave Manchester Parish for high school. It pained him to see neighborhood boys and girls help other schools defeat Manchester High at Champs. In an effort to hoist his local team back atop Champs, Fuller began to steer local runners to Manchester High. Runners like Sherone Simpson.

In '97, Fuller saw Simpson run in a local 100-meter race for twelve-year-olds. His mellow, baritone voice rumbles when he describes it. "She ran 12.2 seconds, hand-timed," he says, his eyes widening. "And that was bare feet, in the grass!" Fuller marveled at Simpson's lithe build. It reminded him of Grace Jackson, a Jamaican Olympian of the 1980s.

But Simpson was an excellent student, and her primary school exam scores had earned her placement at Knox College, a premier academic school in Jamaica, and one without a track team. So Fuller intervened.

He convinced Simpson's parents, Audley and Vivienne, of their daughter's potential on the track. Once they agreed, Fuller got Manchester principal Branford Gayle on board. Gayle contacted Knox

College, and, after some prodding, Knox agreed to grant Simpson a transfer.

For the first few years, she ran well at Champs, but Simpson was more focused on school. High school coaches in Jamaica are generally very conservative in training—most underclassmen don't practice every day, and athletes don't lift weights until at least fifteen or sixteen. High school practice, Simpson says, "was not intense."

But in 2003, her last year at Manchester, Simpson blossomed. She finished second in the 100 at Champs by a shoulder blade to future Olympic medalist Kerron Stewart. Scouts from American colleges, marked by shirts and hats with matching logos, prowl the stands at Champs. (Some scouts also stand out by virtue of the small number of white spectators in the stadium. When I visited Champs, a teenage boy approached me, uttering, "Excuse me, sir?" several times before I realized he meant me. "Do you have any scholarships available?" I was sorry to disappoint him.) Simpson was on the verge of accepting a full scholarship to the University of Texas–El Paso when one of her track-and-field guardian angels intervened, again.

Nearby at UTech—where Errol Morrison is president—coach Stephen Francis was busily molding the MVP Track Club in an effort to give Jamaican athletes a venue to continue training after high school without leaving for the United States and the NCAA track system that Jamaican coaches feel over-races athletes. Manchester High's principal Gayle called Simpson into his office: " 'You'll do UTech for a year, and see how it is,' " Gayle recalls saying. "Then I let her cry, and wipe her eyes. And then she agreed."

In 2004, as a freshman at UTech, Simpson exploded on the international scene, finishing sixth in the finals of the 100-meters at the Athens Olympics. A week later, and just two weeks after her twentieth birthday, Simpson ran down U.S. superstar Marion Jones on the second leg of the 4×100-meter relay and became the youngest gold medalist in Jamaican history. Four years after that, in Beijing, Simpson tied for the 100-meters silver medal, behind UTech classmate Shelly-Ann

Fraser-Pryce, and tied to the hundredth of a second with Kerron Stewart, who had nipped her at Champs five years earlier. Jamaica, 1-2-2 on the Olympic podium.

On a sweltering spring day, reclining on a concrete bench in view of the majestic Blue Mountains and beside the undersized grass track where the MVP Track Club trains, Simpson's lips curl up toward her impossibly high cheekbones when she thinks back on her journey. "I remember it vividly, when Mr. Fuller saw me race the first time, and he came and told me I have a lot of potential," she says. "It all started from there!"

Simpson's story is emblematic of the best of the Jamaican system: nearly every kid is made to sprint at some point in youth races (Simpson's first wins came as a five-year-old in relays at the annual sports day held for Jamaican schoolkids), and adult track enthusiasts, like Fuller and Gayle, keep their eyes peeled for speedy youth and recruit them to good track high schools. There, they are developed very slowly, but get big-race experience at Champs, where they earn adoration and scholarships by performing well. Or, for the best of the best, a shoe company endorsement and membership in a pro club.

The Jamaican sprint system resembles football in the United States, replete with its own shady boosters. (Several high school coaches at Champs told me that they are now banned from giving refrigerators to parents in an effort to recruit their children.) This island-wide, sprint-talent-spotting-and-capture system has paid off in Olympic gold for Jamaica. None other than Usain Bolt pined to be a cricket star in his youth (his second choice was soccer) until he started blowing away his peers in sprints on sports day and was pushed into track and field as a fourteen-year-old—and even then was renowned for ditching practice—ultimately setting Champs records in the 200 and 400 in 2003. Yohan Blake, Bolt's training partner who finished second to him in the 100 and 200 at the 2012 London Olympics, also wanted to be a cricketer, but was identified as a sprinter during sports day at age twelve. Even top American sprinters often come via the Jamaican talent-spotting system. Sanya Richards-Ross, an American who won gold in the 400-meters in London, lived in Jamaica until she was twelve and was plucked by a primary school track coach when, at seven years old, she outstripped older girls in races on sports day. "The coach said, 'Yep, you're coming out for the track team,'" Richards-Ross says.

Physiology findings indicate that endurance training can enhance the ability of fast-twitch muscle fibers to resist fatigue, but that sprint training does not increase the speed at which slow-twitch fibers contract. So being endowed with a large proportion of fast-twitch fibers is essential for an elite sprinter. Or, in the dogma of football coaches: "You can't teach speed." This is an exaggeration, as speed—and certainly the ability to sustain speed—can be improved. But recall the Netherlands' Groningen soccer talent studies. No matter the training, the slow kids never catch up to the fast kids in sprint speed. And the words of Justin Durandt, manager of the Discovery High Performance Centre at the Sports Science Institute of South Africa: "We've tested over ten thousand boys, and I've never seen a boy who was slow become fast." Slow kids *never* make fast adults. So keeping the swiftest kids in the sprint pipeline is paramount. And in what country other than Jamaica could a boy with blinding speed and who stands 6'4" at the age of fifteen, as Bolt did, end up anywhere but on the basketball or volleyball court or the football field? If he's born in the United States, Bolt is no doubt ushered toward the path of towering speedsters like Randy Moss (6'4") and Calvin Johnson (6'5"), both large, fast NFL wide receivers who made many millions of dollars. (Johnson's size and speed helped him land a $132 million contract in 2012.)

The sprint results at Champs are actually comparable to those at state championship meets in big sprinting states, like Texas, and the Champs atmosphere has its fervor in common with Texas high school football. But scores of America's would-be Olympic sprinters land instead in sports that are more popular in the United States, like basketball and football. (A Jamaican sportswriter I met at Champs was concerned that the rising popularity of basketball on the island could siphon off track talent.)

Trindon Holliday, an NFL wide receiver, was such an outstanding sprinter at Louisiana State University that he beat Florida State's Walter Dix—who would take bronze behind Bolt in Beijing—in the 100-meters at the 2007 U.S. national championships, but subsequently gave up his spot on the U.S. world championship team so that he wouldn't miss a day of preseason practice for LSU football. Xavier Carter, who was at LSU at the same time as Holliday, chose to go pro as a sprinter only after failing to make an impact as a wide receiver in two years with the football team. In Jamaica, a key to world sprint domination is keeping the best sprinters on the track.

It is the island-wide talent-spotting system—in which every kid is made to try sprinting at some point—that Pitsiladis credits with Jamaican sprint success. Not to say that genes don't matter. "You absolutely must choose your parents correctly to be a world record holder," he says, rhetorically. "But Jamaica has thousands and thousands sprinting, and you get the best coming through. That's what accounts for this phenomenon. If you had this in any other country, you would see exactly the same thing."

When a Scottish publication solicited Pitsiladis's advice for aspiring United Kingdom athletes, he responded: "Go into sprinting. Don't worry because you're white. It's got nothing to do with the color of your skin."

His friend and colleague Errol Morrison would heartily disagree.

Malaria and Muscle Fibers

Compared with Europeans, Jamaicans have longer legs relative to body height, and more narrow hips. This, Morrison says, is inarguable.

That Jamaicans would have a more linear build than Europeans is no surprise, nor is it specific to Jamaicans. As Allen's rule of body proportions dictates, men and women with recent ancestry from low latitudes and warm climates generally have proportionally long limbs. Another ecogeographic principle, known as Bergmann's rule—named for nineteenth-century biologist Carl Bergmann—indicates that humans with recent low latitude ancestry will also tend to be more narrow, with slimmer pelvic bones. Both long legs and narrow hips are advantageous for running and jumping. All other factors being equal, maximum running speed scales with the square root of leg length. But the theory of western African sprint dominance that Morrison coauthored is a thesis entirely apart from these anatomical concerns.

In 2006, Morrison, with Patrick Cooper, proposed in the *West Indian Medical Journal* that rampant malaria along the west coast of Africa, from where slaves were taken, led to specific genetic and metabolic alterations beneficial for sprint and power sports. The hypothesis: that malaria in western Africa forced the proliferation of genes that protect against it, and that those genes, which reduce an individual's ability to make energy aerobically, led to a shift to more fast-twitch muscle fibers, which are less dependent upon oxygen for energy production. Morrison helped with the biology details, but the fundamental idea originally came from Cooper, a writer and childhood friend of Morrison's.

Cooper was a polymath who had professional success in jobs ranging from music recording to writing speeches for Norman Manley, an architect of Jamaica's independence, and then for his son, Prime Minister Michael Manley. Early in his career, Cooper had been a reporter for *The Gleaner*, Jamaica's largest newspaper. Working at *The Gleaner*'s sports desk, he first surmised that white athletes had historically dominated sprint and power sports only by systematically excluding or dodging black athletes, like boxing champion Jack Johnson. In later writing, Cooper meticulously documented the fact that athletes with western African heritage become highly overrepresented in sprint and power sports almost immediately once they are allowed a fraction of their white counterparts' access to sports. Cooper highlighted trends that continue today: At every Olympics after the U.S. boycott of 1980, every single finalist in the men's Olympic 100-meters, despite homelands that span from Canada to the Netherlands, Portugal, and Nigeria, has his recent ancestry in sub-Saharan West Africa. (The same has been true for women at the last two Olympics, and all but one female winner since the U.S.-boycotted 1980 Games has been of recent western African descent.) And there has not been a white NFL player at cornerback, football's speediest position, in more than a decade.*

As a speechwriter during Michael Manley's combative 1976 reelec-

---

*There are white men in the NFL who play safety—the other defensive back position—and some writers, most prominently William C. Rhoden of the *New York Times*, have argued that would-be white cornerbacks are stereotyped as slow and are shuffled to the safety position by narrow-minded coaches. Stereotyping may be a contributing factor, but data from the NFL's predraft combine also shows that safeties, no matter their ethnicity, perform worse on speed and quickness tests than cornerbacks. As Heisman Trophy winner and Redskins quarterback Robert Griffin III put it on ESPN: "Safeties play safety for a reason, they're not fast. They're not as fast as corners, I should say." A 2011 study in the *Journal of Strength and Conditioning Research* concluded, "cornerbacks would overall appear to be the most, whereas offensive guards would seem to be the least, athletic of the 15 positions examined."

tion campaign, Cooper and his family were under constant threat. Cooper stopped sitting with his back to windows, and when his wife, Juin, was held up at gunpoint, he moved the family away from Jamaica, for good. Living in Houston in the late 1980s, Cooper haunted the library, stalking historical and biological explanations for the dominance of black athletes in sprint sports. Cooper read voraciously from scientific publications in biology, medicine, anthropology, and history in a manner that few ever did prior to the advent of electronic databases that sift scholarly journals with a keystroke.

Cooper found the famous body types study of 1968 Olympians, and he latched on to a curious side note recorded by the scientists. The researchers had been surprised to find that "a sizeable number of Negroid Olympic athletes manifested the sickle-cell trait." That is, some black Olympians had, in one of two copies of the gene that codes for hemoglobin—the oxygen-carrying molecule in red blood cells—a mutation that causes round red blood cells to curl up in a sickle shape in the absence of oxygen, potentially impairing blood flow through the body during vigorous exercise. The gene variant that causes sickle-cell trait is found most often in people with recent sub-Saharan ancestry in west or central Africa, and scientists had previously believed that the high altitude of the 1968 Mexico City Olympics would prevent athletes with sickle-cell trait from performing well. "Sickle-cell was supposed to be a deterrent," Morrison says. But it made no difference at the Olympics in events of short duration, like sprints and jumps.

In the decades since, epidemiological studies have found that athletes with sickle-cell trait (they have one copy of the mutant gene and are known as "sickle-cell carriers") are indeed underrepresented in athletic endeavors that require aerobic endurance. In competitive running, sickle-cell carriers all but disappear in events longer than 800 meters. They are genetically disadvantaged for long-distance sports. In a small number of sickle-cell carriers, blood flow is inhibited to such a degree as to become deadly if they work out too hard for too long. Since 2000, the sudden deaths of nine college football players—all of them black and in Division I—during training have been tied to sickle-cell trait, and the NCAA now requires screening for the gene variant that causes it. (According to a panel at the 2012 Big East Conference Sports Medicine Society, white college athletes, on the advice of a team doctor, will often sign a waiver to forgo the testing, given the unlikelihood that they carry the sickle-cell gene variant.)

In 1975, the year after the Mexico City Olympics data was published, another study appeared that Cooper would dissect two decades later, this one showing naturally low hemoglobin levels in African Americans. The work was published in the *Journal of the National Medical Association*, run by the Maryland-based National Medical Association, which promotes the interests of physicians and patients of recent African descent. Using data from nearly 30,000 people in ten different states, with ages ranging from the first year to the ninth decade, it reported that African Americans have lower hemoglobin levels at every stage of life than white Americans, even when socioeconomic status and diet are matched. (Errol Morrison's wife, Fay Whitbourne, formerly head of Jamaica's National Public Health Laboratory Services, says that hemoglobin levels among Jamaicans are in line with those of African Americans.) Numerous studies, as well as population data from the U.S. National Center for Health Statistics, have replicated this result in the years since, including in athletes. In a colossal 2010 study of 715,000 blood donors across America, researchers wrote that African Americans exhibit a "lower genetic set point for hemoglobin," regardless of environmental factors like nutrition.* Like sickle-cell trait, genetically low hemoglobin—all else being equal—is a genetic *disadvantage* for endurance sports. Runners of recent western African descent are very much underrepresented at high levels of distance running. (The Jamaican record in the 10K would not even have qualified for the 2012 Olympics)

The authors of the *Journal of the National Medical Association* paper

---

*The authors noted that black blood donors are sometimes inappropriately turned away because their low hemoglobin levels are assumed to be the result of a health condition.

wrote that lower hemoglobin levels raise the possibility that African Americans employ more of some alternate energy pathway to compensate for a relative lack of oxygen-carrying hemoglobin. Two years later, in the same journal, another group of scientists insisted: "some compensatory mechanism must exist to counteract this relative deficiency in hemoglobin, since a significant difference has even been demonstrated in healthy athletes." Cooper set out to find that compensatory mechanism.

His tireless perusal of medical journals took on greater urgency in 1996, when he was diagnosed with terminal prostate cancer. Cooper and Juin moved to New York City in 2000 so that Cooper could spend every day at the New York Public Library. "My office," he called it. Weekend trips to Baltimore to visit his daughter doubled as visits to the University of Maryland library.

And then Cooper found just the potential "compensatory mechanism" he was looking for, in a 1986 study from Laval University in Quebec published in the *Journal of Applied Physiology* and coauthored by Claude Bouchard, who would go on to become the most influential figure in the field of exercise genetics, and the leader of the HERITAGE Family Study that documented aerobic trainability differences among families. Bouchard and colleagues took muscle samples from the thighs of two dozen sedentary Laval students, primarily from countries in western Africa, as well as from two dozen sedentary white students, who were identical to the African students in age, height, and weight. The researchers reported that a higher proportion of muscle in the African students was composed of fast-twitch muscle fibers, and a lower proportion was slow-twitch muscle fibers compared with the white students. The African students also had significantly higher activity in the metabolic pathways that rely less on oxygen to create energy and that are engaged during an all-out sprint. The scientists concluded that, relative to the white students, the students from western Africa "are, in terms of skeletal muscle characteristics, well endowed for sport events of short duration."

The study was small, as usual with biopsy studies that require the surgical removal of a gobbet of muscle tissue. The few similar studies over the years have generally agreed with the Laval findings, but each one has relied on a small number of subjects.*

In his 2003 book, *Black Superman: A Cultural and Biological History of the People Who Became the World's Greatest Athletes*, and then in his 2006 paper with Morrison, Cooper first made the argument that West Africans evolved characteristics like a high prevalence of the sickle-cell gene mutation and other gene mutations that cause low hemoglobin for protection from malaria, and that an increase in fast-twitch muscle fibers followed from that, providing more energy production from a pathway that *does not* rely primarily on oxygen, for people who have reduced capacity to produce energy *with* oxygen. The former part of Cooper's hypothesis—that sickle-cell trait and low hemoglobin are evolutionary adaptations to malaria—now seems undeniable.

In 1954, the same year Sir Roger Bannister broke the four-minute mile, British physician and biochemist Anthony C. Allison, who had been raised on a farm in Kenya, showed that sub-Saharan Africans with sickle-cell trait have far fewer malaria parasites in their blood than inhabitants of the same region who do not have sickle-cell trait. Normally, the sickle-cell gene variant seems like a bad thing to carry. If two people who each have one copy have kids together, one in four of their children will have two copies of the gene and therefore sickle-cell *disease*—also known as sickle-cell anemia—a condition in which sickled blood cells exist even without exercise, and life expectancy is reduced. And yet, this gene mutation has hung

---

*Says Bouchard of the finding that subjects with recent West African ancestry had more fast-twitch muscle fibers: "They had a bit more type II [fast-twitch] muscle fibers. Not a difference in kind, a difference in the frequency of events, which means there would be more people with the basic biology who, if selected and trained, might achieve success more readily than the average person of European ancestry. But we do have people of European ancestry with the same profile. That was our conclusion and I have seen no data to make me change my mind." Bouchard also noted that a small difference in the average means a large difference in people at the end of the curve who have extreme biology.

around—proliferated, actually—in the malaria danger zones of sub-Saharan Africa.

That is because people who have one copy of the sickle-cell gene variant are generally healthy, but have red blood cells that sickle when infected with the malaria parasite, which in turn protects the host from the parasite's devastating effects. (Because sickle-cell disease shortens lives, the sickle-cell gene will never spread through an entire population. Among African Americans who have lived in the malaria-free United States for generations, the sickle-cell gene variant is steadily disappearing.) Today, the sickle-cell balance with malaria resistance is one of biology's textbook examples of an evolutionary tradeoff, propagating an otherwise harmful gene variant because of an associated protection.

Cooper and Morrison's suggestion that low hemoglobin in African Americans and Afro-Caribbeans is a second adaptation to malaria has been proven true as well, in a deadly manner.

Even as evidence mounted that low hemoglobin levels in Africans native to malarial zones is at least partly genetic, aid workers in Africa looked upon low hemoglobin as a sign purely of a diet with too little iron. In 2001, the United Nations General Assembly charged the world with reducing iron deficiency among children in developing nations. And so, in a well-intended effort to improve nutrition, health-care providers descended on Africa with iron supplements, which raise the hemoglobin levels of those who consume them. (Hemoglobin is an iron-rich protein, so levels fall if insufficient iron is consumed. Often the first thing elite endurance athletes check for if they start performing poorly is a low iron level.)

The problem was that doctors who studied malarial regions saw increased cases of severe malaria wherever iron supplements were dispensed. Since the 1980s, scientists working in Africa and Asia had documented lower rates of malaria death in people with low hemoglobin levels. In 2006, following a large, randomized, placebo-controlled study in Zanzibar that reported a stark increase in malaria illness and death among children given iron supplements, the World Health Organization issued a statement backtracking from the earlier UN position and cautioning health workers about giving iron supplements in areas with high malaria risk. Low hemoglobin, like sickle-cell trait, is apparently protective against malaria. And, in keeping with Cooper and Morrison's hypothesis, many Africans who were forcibly taken to the Caribbean and North America came from the precise parts of the west coast of sub-Saharan Africa that suffer the highest rates of malaria illness and death in the world, as well as the greatest frequency of the sickle-cell gene.

It is the coda of the Cooper and Morrison hypothesis—that fast-twitch muscle fibers moved in as hemoglobin moved out—that is highly speculative.

To the end of his life, Patrick Cooper remained dedicated to his research and writing. Up until the day in 2009 that cancer finally overwhelmed him, Cooper was dictating to Juin from his bed. I had been hoping to meet Cooper on my trip to Jamaica before I learned that he had passed away and hadn't been living in Jamaica for years anyway. Instead, I met with Morrison and then presented the paper he and Cooper coauthored to five scientists who were not previously familiar with it, and asked their opinions. One insisted that the theory was too speculative to discuss. The other four said that it was a reasonably constructed hypothesis, but also that it had never been directly tested and was not proven. (In 2011, though, scientists from the University of Copenhagen proposed that a high proportion of fast-twitch muscle fibers could account for several physical traits that have been documented in African Americans and Afro-Caribbeans, including low resting and sleeping metabolism, and less metabolism of fat for energy and more of carbohydrates as compared with Europeans.)

Pitsiladis—the gene hunter who collects DNA from world-class sprinters—argues that such a theory could not hold true because of the tremendously diverse genetic background of African Americans and Jamaicans that shows they aren't some genetically monolithic block.

But they do have the traits in question—significant prevalence of sickle-cell trait and low average hemoglobin—in common, so the issue of general genetic diversity is irrelevant. Africans are, on average, much more genetically diverse than Europeans. But with respect to certain genes, like the ACTN3 sprint gene variant, they can be more homogenous. So genetic diversity in itself does not imply that an ethnic group cannot share a common trait, as many certainly do. As Yale geneticist Kenneth Kidd said of African Pygmy groups: they are among the most genetically diverse people in the world, and yet they share the trait of diminutive stature that will prevent them from dominating the NBA.

Because I could not follow up with Cooper himself, I decided to follow up on his work to see if any evidence had emerged that might affirm or dismantle his theory since it was published. First stop: do athletes with sickle-cell trait perform any differently in explosive sports?

French physiologist Daniel Le Gallais, former medical director of the National Center for Sports Medicine in Abidjan, Ivory Coast, posed that question long before Cooper. About 12 percent of Ivorian citizens are sickle-cell carriers, and in the early 1980s Le Gallais noticed that the top three female Ivorian high jumpers (one of whom won the African championship) became abnormally exhausted during workouts. Le Gallais tested the athletes and found—"surprisingly," he wrote in an e-mail—"these three athletes were sickle cell trait carriers, despite originating from different ethnic groups in the country."

Le Gallais later coauthored studies that screened for sickle-cell trait in elite sprinters and jumpers. In 1998, he reported that nearly 30 percent of 122 Ivorian national champions in explosive jumping and throwing events were sickle-cell trait carriers, and that they collectively accounted for thirty-seven national records. The top male and female in the group were both sickle-cell carriers. In a 2005 study of sprinters from the French West Indies who made the French national team, about 19 percent of the athletes tested were sickle-cell carriers, and they accounted for an outsized proportion of titles and records held by the team.

"What is my standpoint currently?" Le Gallais wrote me. "Studies have clearly shown that athletes with [sickle-cell trait] were less numerous than non-SCT athletes in long endurance races. In contrast, athletes with SCT are more numerous in jumps and throws. . . . The oxygen transport system impairment explains the poor performances in long distance races. On the contrary, we don't know the cause of their advantage in jumps and throws."

As for whether low hemoglobin in itself might prompt a switch to more fast-twitch fibers, there is evidence that it can in rodents. A UCLA study of mice that were put on iron-deficient diets showed a drop in hemoglobin and displayed a shift of type IIa fast-twitch muscle fibers to type IIb "super fast twitch" muscle fibers in their lower legs. In another study in Spain, rats were made to have low hemoglobin through periodic blood draws, and a shift to a higher proportion of fast-twitch fibers occurred in their lower legs. But no one has conducted such a study in humans, and mice have a greater ability to swap muscle fiber types than humans do. Plus, that is a developmental effect within the lifetime of a mouse, not an evolutionary one caused over generations by changing genes.

And that is all the science there is. A single mouse study and a single rat study demonstrating in rodents that low hemoglobin can induce a switch to more explosive muscle fibers. No scientist has attempted to test Cooper and Morrison's idea in humans, so there are simply no human studies at all.

Several scientists I spoke with about the theory insisted that they would have no interest in investigating it because of the inevitably thorny issue of race involved. One of them told me that he actually has data on ethnic differences with respect to a particular physiological trait, but that he would never publish the data because of the potential controversy. Another told me he would worry about following Cooper and Morrison's line of inquiry because any suggestion of a physical advantage among a group of people could be equated to a corresponding lack of intellect, as if athleticism and intelligence were on some kind of biological teeter-totter. With that stigma in mind, perhaps the most important writing Cooper did in *Black Superman* was his methodical evisceration of any supposed inverse link between physical and mental prowess. "The concept that physical superiority could somehow be a symptom of intellectual inferiority only developed when physical superiority became associated with African Americans," Cooper wrote. "That association did not begin until about 1936." The idea that athleticism was suddenly inversely proportional to intellect was never a cause of bigotry, but rather a result of it. And Cooper implied that more serious scientific inquiry into difficult issues, not less, is the appropriate path.

Cooper and Morrison's hypothesis, that reduced oxygen-carrying capacity induced a shift to more explosive muscle properties, was never intended as simply a "black" phenomenon. Even if the hypothesis is correct, there is still tremendous physiological variation within any ethnic group, and Cooper and Morrison were theorizing about a set of black athletes with very specific geographic ancestry.

On the side of Africa opposite the ancestry of sprinters, and by the serendipity of geography, a different faction of the world's greatest athletes were spared potentially endurance-harming genetic adaptations. They live at altitudes where the mosquitoes are scarce, and so are malaria and the sickle-cell gene.

Those black athletes came to dominate in an entirely different realm.

12

Can Every Kalenjin Run?

Every summer, John Manners returns to Kenya, and every July—after the 1,500-meter time trial—there are tears. Most of them stream down the cheeks of the kids who just ran. But, says Manners, "some of the tears are mine. It's a pretty emotional business."

It's hard to imagine Manners sad. His eyes glitter under a newsboy cap. Together with his pointed white goatee and his buoyant walking stride, the eyes lend a puckish delight to his conversations.

The 1,500-meter race that makes Manners cry is the capstone of a unique college application process for sixty or so impoverished Kenyan kids each year, and Manners and his KenSAP program have to leave all but a dozen of them behind.

Begun in 2004, KenSAP—the Kenya Scholar-Athlete Project—is the brainchild of Manners, a New Jersey–based writer, and Dr. Mike Boit, a bronze medalist for Kenya in the 800-meters at the 1972 Olympics and now a professor of exercise and sports science at Kenyatta University in Nairobi. The idea is to get top Kenyan students from the western Rift Valley Province into premier colleges in the United States.

Each year, Manners peruses the list in the newspaper of the highest scorers on the Kenya Certificate of Secondary Education (KCSE) exam—a high school exit exam that accounts for 100 percent of the college admissions process in Kenya—for names of students with the best marks in the western Rift Valley. He also goes on local Kass FM radio and solicits applications from students who scored an "A plain," the highest possible mark. Still, recruitment has challenges. "Because the program's free," Manners says, "some of the [applicants'] parents assume it's a scam."

Manners invites selected students who complete an application to the High Altitude Training Center, in the Rift Valley town of Iten. There they are interviewed, and then made to run a 1,500-meter race at an altitude around 7,500 feet. All of the students have succeeded in high school despite coming from destitute rural families. The majority are boys—the patriarchal nature of Kenyan culture affords girls less opportunity to prepare for the KCSE exam—and some come from tiny subsistence farms and attend school in classrooms with mud or stone floors. All have both the academic skill and the college-essay fodder to knock the argyle socks off East Coast admissions officers. After the interview and 1,500, Manners confers with Boit and a group of American instructors and local Kenyan elders, and within hours reads aloud the names of the kids who are accepted. That's where the tears come in, from those who missed the cut.

The dozen kids KenSAP accepts undertake two months of intensive SAT prep and college application work. Thus far, the KenSAP plan has worked brilliantly. Between 2004 and 2011, seventy-one of the seventy-five students accepted by KenSAP gained entrance to U.S. colleges. Every Ivy League university has had a KenSAP kid. Harvard leads the league with ten, followed by Yale at seven, and Penn with five. Others have gone to prestigious liberal arts colleges, on the order of Amherst, Wesleyan, and Williams. "I love NESCAC," says Manners, referring to the New England Small College Athletic Conference. "We're very strong in NESCAC."

The 1,500-meter time trial is, obviously, an unprecedented piece of a college application process. Kenyan kids who score an A plain usually come out of government-supported boarding schools, and most have no running experience at all. In a letter sent to KenSAP applicants months before the interviews, Manners explains that there will be a running test, and that they should dress accordingly. And yet, without fail, some boys will show up in long pants, and a few girls in calf-length skirts and pumps.

Manners's hope with the 1,500 is to find undiscovered athletic prodigies with the running chops that will persuade an American coach to put a word in with the admissions committee. "We're looking for everything we can to strengthen an application," Manners says. If a kid with no running background shows promise, Manners will contact college coaches to see if any might be interested.

If forcing the academic all-stars of a geographic sliver of East Africa to run a 1,500-meter time trial on a dirt track at 7,500 feet seems a little strange, well, it is. Imagine a college admissions counselor taking the American kids who scored a perfect 2400 on the SAT and lining them up for a time trial.

But then, this is no random geographic sliver.

In 1957, when Manners was twelve, he moved with his father from Newton, Massachusetts, to Africa. Robert Manners, an anthropology professor and founder of the anthropology department at Brandeis University, had intended to study the Chaga people of Tanzania. But another anthropologist beat him to it, so Manners ventured west to the Rift Valley of Kenya to study the Kipsigis, a traditionally pastoral people who are a subgroup of a larger tribe, the Kalenjin. The Kipsigis held fiercely to their traditional culture in the face of British colonization, which lasted until 1963.

Robert Manners found a house in Sotik, in western Kenya, surrounded by tea and cattle farms, and at an altitude of six thousand feet. There was one mud street, enclosed by verandas over raised sidewalks, like a town from the Old West. In short order, John Manners became like any other Kipsigis child, speaking Swahili and running two to three miles to school with his friends so they could avoid being caned for showing up late. He also attended his first track meet, as a spectator.

As was the case in Jamaica, British colonialism imported the sport of track and field. The Kenya Amateur Athletics Association was founded in 1951, and by the time the Manners family arrived, regional track meets—on dirt or grass tracks—were common. At one of the first meets Manners saw, in seventh grade, he was delighted by the stellar performances of Kipsigis runners—*his people.*

In the fall of 1958, Manners returned to Massachusetts for eighth grade, but his fascination with track and field, and with Kenya, remained. In the 1964 Olympics, just the third ever in which Kenya competed, a Kipsigis runner named Wilson Kiprugut won bronze in the 800-meters. Four years later, in the altitude of Mexico City, Kenya was the dominant distance running power, winning seven medals in middle- and long-distance events. The very same month of those Olympics, Manners, having just graduated from Harvard, was in up-state New York training for the Peace Corps. "I saw the names of the Kenyan runners who were winning those medals," Manners says, "and I saw that almost all of them were Kalenjin."

Manners was exhilarated by the success of Kenyan runners, as it defied stereotypes held by British colonialists. "The conventional wisdom was that blacks could sprint, but that anything that required tactical sophistication, or discipline, or training," he says, "this was the white man's province."

With the Peace Corps, Manners returned for another three years to the western Rift Valley in Kenya, where locals still remembered him and his father. In the early 1970s, a few Kenyan middle- and long-distance runners began to show up on American college campuses, and Manners started writing about Kenyan running. In 1972, he co-authored an article for *Track & Field News*: "Basically, the piece said that coaches in America are wondering whether there are more great runners back there in Kenya," Manners says. "And our answer was: Thousands!" Particularly among the Kalenjin.

The 4.9 million Kalenjin people represent about 12 percent of Kenya's population, but more than three quarters of the country's top runners. In 1975, in a footnote to a chapter he contributed to *The African Running Revolution*, a book compiled by *Runner's World* magazine, Manners raised an evolutionary theory of Kenyan—and specifically Kalenjin—running success that remains controversial today.

Manners wrote that a part of traditional life for Kalenjin warriors was the practice of cattle raiding. Essentially, it entailed stealthily running and walking into the land of neighboring tribes, rounding up cattle, and escorting them back to Kalenjin land as quickly as possible. Cattle raiding was not considered theft so long as the raiders weren't filching the cattle from the same subtribe within the Kalenjin. "The raids were conducted largely at night," Manners wrote, "and sometimes ranged over distances as great as 100 miles! Most raiding parties were group ventures but each *muren* [or warrior] was expected to at least do his share."

A *muren* who brought back a large number of cattle from a raid was hailed as a courageous and athletic warrior and could use his cattle and prestige to acquire wives. In a footnote, Manners wrote that, insofar as successful cattle raiders had to be strong runners to hustle captive herds to safety, and the best cattle raiders accumulated more wives and children, then cattle raiding could serve as a mechanism of reproductive advantage that favored men with superior distance running genes. In the next breath of the very same chapter, though, Manners seems to doubt the suggestion as soon as he raises it. "The idea just occurred to me, so I just put it in," he says now.

But over the years, as he has continued to study Kalenjin running, and to interview Kalenjin runners and elders, he has come to regard the idea as much less fanciful—in part because other "hot spots" of endurance running talent have materialized in East Africa, and the athletes responsible are also from traditionally pastoralist cultures that once practiced cattle raiding.

In Ethiopia, the world's second distance running superpower, the

Oromo people make up about one third of the country's population but the vast majority of its international runners. The Sebei people of Uganda—who live just across Mount Elgon from Kenya—account for essentially all of that nation's top distance runners and include Stephen Kiprotich, who won the 2012 London Olympics marathon. The Ugandan Sebei are actually a subgroup of Kenya's Kalenjin.

In a converted attic storage room, under the slope of the roof on the third floor of his house in Montclair, New Jersey, Manners has his office. It's the kind of eruption of paper and maps that one might find as the parent of a brilliant twelve-year-old who has been quietly making plans to visit Mars. Files, books in stacks, books on shelves, maps. Giant maps, affixed to the slanted ceiling, dotted with meaningfully placed tacks.

The maps show the specific districts of western Kenya from which runners flow forth en masse. Beside the maps sit every *Association of Track and Field Statisticians Annual* published since 1955. The ATFS is a volunteer group of track stats junkies, and many of the *Annual*s are long out of print. "I had to buy some of them from collectors," Manners says. He also has nearly every *African Athletics* annual ever published, as well as a complete collection of *Track & Field News* dating back to 1971.

Manners has catalogued the specific geographical distribution and tribal membership of Kenyan runners—often by asking the runners in person—to a greater extent than any other human being alive. Along the way, he has collected staggering anecdotes of gifted Kalenjin runners.

Like the one about Amos Korir, who was supposed to compete in pole vault at the Community College of Allegheny County in Pennsylvania when he arrived there in 1977. But upon seeing how much better the other vaulters were he fibbed to the coach, claiming to be a runner. Korir was thrust into the 3,000-meter steeplechase—a race just shy of two miles that includes hurdles—and in his third-ever attempt at the event won the national junior college championship. Four years later, Korir was the third-ranked steeplechase runner in the world.

Or the one about Julius Randich, who arrived at Lubbock Christian University in Texas a heavy smoker with no competitive running background. By the end of his first year, 1991–92, Randich was the national small-colleges (NAIA) champion in the 10K. The following year, Randich set NAIA records in the 5K and 10K and was named the outstanding athlete in any sport in the NAIA. Kalenjin runners became all the rage among NAIA coaches, and several others would win the 10K national championships after Randich, including his younger brother Aron Rono, who won it four straight times.

And then there's the one about Paul Rotich, perhaps the most famous of Manners's anecdote collection. Rotich, the son of a prosperous Kalenjin farmer, arrived at South Plains Junior College in Texas in 1988, having lived a "comfortably sedentary" life, as Manners describes it. Rotich, a stout 5'8" and 190 pounds, quickly burned through most of the $10,000 his father had given him for two years of living expenses and tuition. "But rather than return home in disgrace," Manners wrote, "Paul . . . decided to train in hopes of earning a track scholarship." Rotich trained at night to avoid the embarrassment of being seen. That concern would be short-lived, as he made the national junior college cross-country championships in his first season. He went on to become a ten-time All-American in cross-country and indoor and outdoor track. As Manners reported, when Rotich returned to Kenya and detailed his running exploits to a cousin, the cousin replied: "So, it is true. If you can run, any Kalenjin can run."

Manners does not think that *any* Kalenjin can be a great distance runner, but he does believe that the proportion of people who will become extremely fast middle- and long-distance runners extremely quickly upon training is significantly higher among the Kalenjin than it is among other tribes in Kenya, or among other peoples throughout the world.

Consider this: seventeen American men in history have run a marathon faster than 2:10 (or a 4:58 per mile pace); thirty-two Kalenjin men did it just in October 2011.* The statistics that describe Kalenjin distance running dominance are endless, and often so outlandish as to be laughable. For example: five American high-schoolers have run under four minutes in the mile in history; St. Patrick's High School, in the Kalenjin training town of Iten, once had four sub–four milers in school at the same time. (Conversely, the Kenyan record in the 100-meters, 10.26 seconds, wouldn't even have made the bare minimum standard to participate in the London Olympics.) Wilson Kipketer, a former St. Patrick's student who became a Danish citizen and held the 800-meter world record from 1997 to 2010, does not hold his own high school's record. (That distinction belongs to Japhet Kimutai, who ran 1:43.64.)

Manners was banking on the western Rift Valley's fountain of talent when, in 2005, he held KenSAP's first "great tryout," as he calls it. While scientists and running enthusiasts have assayed Kenyan dominance every which way to make points about whether or not Kenyan runners are genetically gifted for endurance running, Manners's tryout—which has the goal of helping poor Kenyan kids get into elite colleges—is more truly a random sample of Kalenjin than nearly any scientist has ever taken and put on the track. The kids in his time trials generally come from elite, highly selective, government-funded boarding schools, and essentially none of them have any racing experience. This is panning for endurance gold in its most raw form.†

Each year, about half of the boys in the time trial will run faster than 5 minutes and 20 seconds in the 1,500-meter time trial, on a shoddy dirt track, above seven thousand feet. (The 1,500 is about 100 meters shy of a mile, and 5:20 translates to a mile time just over 5:40.)

---

*Seventeen Ethiopians and Kenyans broke 2:10 in a single race in 2012, the Standard Chartered Dubai Marathon.
†As Manners says, he is actually picking *against* finding running skill, because he invites kids "who have spent all their time in high school studying."

"Can you imagine, if you considered a comparable group from any American upper-echelon academic selection?" Manners asks. "I mean, it would be nowhere near that."

In the tryout in 2005, a boy named Peter Kosgei ran 4:15 with no real training. Kosgei was accepted to Hamilton College in Clinton, New York, and quickly became the best athlete in the college's history. In his freshman year, Kosgei won the Division III 3,000-meter steeplechase national title. By the end of his junior year, he had compiled eight more national titles in cross-country and track. His skill was so out of place in Division III that his teammate Scott Bickard compared it to "going to a Division III school to play basketball and you find yourself playing with a guy who can play in the NBA."

Sadly, Kosgei was unable to compete in track during his senior year. On a trip home to Kenya during spring break in March 2011, Kosgei was mugged and left with two broken legs. When I met him at a KenSAP function eight months later, Kosgei was pursuing a graduate degree in chemistry and told me that he aspired to race again one day. At Hamilton, he said, he trained a paltry thirty to thirty-five miles a week, and thus felt that he had only grazed the outermost layer of his potential.

A slew of other KenSAP runners have met quick success. Evans Kosgei—no relation to Peter—held down a 3.8 GPA in computer science and engineering at Lehigh University and, after adjusting to life in America for a year, decided to go out for cross-country in his sophomore year. He struggled even to finish his five-mile tryout. But, in short order, Kosgei was running at the Division I national championships in both cross-country and track. In 2012, he was named Lehigh's Graduating Scholar-Athlete of the Year.

Manners says that many KenSAP students have no interest in running, and some of those who were welcomed by American coaches quickly dropped the sport to focus on academics. But of the seventy-one KenSAP students through 2011—none with significant prior training experience—fourteen made varsity NCAA rosters.

Of course, stumbling upon hidden distance running talent is not exclusive to Kenya. And, as with Jamaican sprinting, it is the very systematizing of the process by which talent is stumbled upon that makes it less like stumbling and more like tactical filtering. The ultimate question is whether finding endurance talent is more likely to occur in Kenya, or specifically among the Kalenjin, and whether that is largely due to innate biological characteristics. For certain sports, it's obvious and uncontroversial that particular populations will have a greater or lesser frequency of gifted prospective athletes. Pygmy populations have an average adult male height of around five feet. So, while they may produce an NBA player someday, a basketball scout taking a random sample from a Pygmy population will discover fewer athletes who, given the proper training, might make the NBA than if the sample were taken in Lithuania.

Presently, there is no way to know how the KenSAP time trial would compare with a similar exercise focused on a different ethnic group in Kenya or somewhere else in the world, and the KenSAP tryout isn't intended to be a scientific experiment. There was one research group, though, that tried to get at the answer in a scientific manner.

Beginning in 1998, a team of researchers from the University of Copenhagen's world-renowned Copenhagen Muscle Research Centre set out to put data behind the many anecdotes and arguments about Kalenjin distance running dominance. Among the theories they sought to investigate: that members of the Kalenjin tribe might have a particularly high proportion of slow-twitch muscle fibers in their legs; that Kalenjin people are born with higher aerobic capacity (VO_2max); and that Kalenjin people might respond more quickly to endurance training than members of other ethnic groups.

To untangle at least a segment of the nature from the nurture, the scientists set out to study not only elite runners, but also Kalenjin boys who lived in cities and those who lived in rural villages, as well as Danish boys living in Copenhagen.

Overall, the findings did not support any of the long-standing but uninvestigated theories. Elite runners from the Kalenjin tribe and from Europe did not differ on average in their proportion of slow-twitch muscle fibers, nor did Danish boys differ from Kalenjin boys who lived in cities or those who lived in rural villages. Kalenjin boys from villages did have higher VO_2max than Kalenjin boys from cities, who were much less active, but it was similar to the VO_2max of the active Danish boys. And Kalenjin boys, as a group, did not on average respond to three months of endurance training—as measured by aerobic capacity—to a greater degree than did Danish boys.

As expected from their latitudes of ancestry, though, the Kalenjin and Danish boys did display body type differences. A greater portion of the body length of the Kalenjin boys was composed of legs. The Kalenjin boys were, on average, two inches shorter than the Danish boys, but had legs that were about three quarters of an inch longer.

The scientists' most unique finding, though, was not the length of the legs, but their girth. The volume and average thickness of the lower legs of the Kalenjin boys was 15 to 17 percent less than in the Danish boys. The finding is substantial because the leg is akin to a pendulum, and the greater the weight at the end of the pendulum, the more energy is required to swing it.* Biologists have demonstrated this in humans in controlled conditions. In one particularly well-controlled study, researchers experimented with adding weights onto different parts of runners' bodies: the waist, the upper thigh, the upper shin, and around the ankle.

Even when the weight stayed the same, the farther down the leg it was placed the greater the energetic cost to the runners. In one phase, each runner had to wear eight pounds around his waist, which required about 4 percent more energy to run at a given pace compared

---

*Oscar Pistorius, the South African double-amputee known as "the blade runner"—who as of this writing is awaiting trial for the murder of his girlfriend—sprints on carbon fiber crescents that are much lighter than human legs. He has the fastest leg swing time ever recorded in a sprinter, by a lot.

with when he wasn't wearing eight pounds of weights. But when the runners were subsequently equipped with a four-pound weight on each ankle they burned energy 24 percent more rapidly while running at the same pace, even though their total weight had not changed one ounce from the previous condition.

Weight that is far out on the limbs is called "distal weight," and the less of it a distance runner has, the better (i.e., if you have thick calves and ankles, you won't be winning the New York City Marathon). A separate research team calculated that adding just one tenth of one pound to the ankle increases oxygen consumption during running by about 1 percent. (Engineers at Adidas replicated that finding in the process of constructing lighter shoes.) Compared with the Danish runners, the Kalenjin runners tested by the Danish scientists had nearly a pound less weight in their lower legs. The scientists calculated the energy savings at 8 percent per kilometer.

"Running economy" is the measure of how much oxygen a runner utilizes to run at a given pace. Much like the fuel economy of a car, you get a certain amount of bang for a certain amount of gas, and that differs according to the size and shape of the car. Elite distance runners have both high VO_2max and good running economy. Or, to continue the car analogy, the rare mix of a big engine and good fuel economy. Among elite runners, all of whom have large engines, running economy often differentiates the extremely great from the merely very good.

And on that measure, untrained Kalenjin boys were better than untrained Danish boys. Proportionally long legs and thin lower legs contribute separately to good running economy, and they had both.[*] Even the Kalenjin city boys, who were less active and less aerobically fit than the Danish boys, started with better running economy. Both

---

[*]A small 2012 study in the *European Journal of Applied Physiology* found that a group of Kenyan runners had Achilles tendons that are 2.7 inches longer than nonrunner white control subjects of the same height. That is to be expected, given the Kenyans' proportionally longer lower limbs. Longer Achilles tendons can store more elastic energy. (Recall: world champion high jumper Donald Thomas.) The next question for scientists: How much do those long tendons influence running ability?

within and between groups of Kenyan and Danish runners, lower leg thickness was an important predictor of running economy. Among Danes and Kenyans who were training similar mileage each week—or not training at all—Kenyans had superior running economy.

That is, when they were using the same proportion of their oxygen-carrying capacity, the Kenyans were going faster for that same effort. As one might expect from the artificial selection for body types that occurs in high-level sports, elite Kenyan runners had even more narrow lower legs—and much better running economy—than did untrained Kenyan boys. One of the scientists, Bengt Saltin, among the most prominent exercise scientists in the world, wrote: "the relationship seems to confirm that the lower leg thickness expressed in absolute terms is a crucial factor for running economy." Later, Henrik Larsen, another researcher in the Copenhagen group, declared: "We have solved the main problem" of Kenyan running dominance.

Lithe legs help running economy no matter one's nationality or ethnicity. One of the best running economies ever measured in a laboratory belonged to Eritrean runner Zersenay Tadese, the world record holder in the half marathon as of this writing. The measurements, taken in a lab in Spain, show that Tadese does not have particularly long legs—his legs are only slightly proportionally longer than those of elite Spanish runners—but they are considerably narrower. Interestingly, Tadese grew up dreaming of a career in competitive cycling—one of the first national sports federations formed in Eritrea was for cycling—but found vastly more success when he switched to running just prior to his twentieth birthday, placing thirtieth at the World Cross Country Championships in his very first season in 2002, before winning the world title in 2007. Surely, Tadese's aerobic fitness from cycling carried over to running, but his thin lower legs are an advantage best exploited on the track, not the bike.

As Tadese proves, it isn't as though thin lower legs are confined to the Kalenjin. But the Kalenjin do, in general, have a particularly linear build, with narrow hips and long, thin limbs. Some anthropologists actually refer to the extreme of a slender body build as the Nilotic type—"Nilotic" refers to a set of related ethnic groups residing in the Nile Valley—and, it so happens, the Kalenjin are a Nilotic people.* The Nilotic body type evolved in low latitude environments that are both hot and dry, because the long, thin proportions are better for cooling. (Conversely, the extreme of the short, stocky build was historically known as the Eskimo type, though the term "Eskimo" has been replaced in some countries, where it is considered derogatory.) And the Kalenjin are as low latitude as it gets. When I visited Kenya in 2012, while driving between training sites I crisscrossed the equator. But the Kalenjin initially migrated to Kenya from southern Sudan, where other Nilotes live today, like the Dinka, an ethnic group known for its tall and slender constituents. A few very long-limbed professional basketball players have been Dinka—most notably Manute Bol, who was 7'7" and reportedly had a wingspan that was 8'6".

Given that the linear build is helpful for endurance running, and that Nilotic people tend to have a linear build, it occurred to me that there should be a wealth of running talent in southern Sudan. But long-distance runners from Sudan are almost absent from international competition. I asked both scientists and track-and-field experts if they had any insight into whether Sudanese runners have been tested for running economy, or why we don't see Nilotic distance runners coming out of Sudan. Unfortunately, there is no data at all on Sudanese runners, and the consensus among track experts was that, unlike Kenya, which apart from postelection violence has been relatively stable, modern Sudan has been in a constant state of tumult and violence that has curtailed opportunities for athletes.

In December 2011, I attended the Arab Games in Qatar and spoke with Sudanese athletes and journalists who told me that, among other

---

*The Kikuyu, the largest ethnic group in Kenya, account for about 17 percent of the population but are somewhat stockier—indicative of their ancestry in a moist, mountainous region—and produce far fewer pro runners than the Kalenjin, who account for just 12 percent of the population. The Kikuyu are a Bantu people.

problems, like travel difficulties, athletes from the southern regions of Sudan (now the nation of South Sudan) had historically been discriminated against and that national sports officers did not enter skilled athletes from that area in past Olympics. Plus, civil war has raged for the better part of a half century in the exact area where the Nilotic people reside, leaving no sports culture or infrastructure whatsoever in southern Sudan. So I approached the question the only way I could think of: looking for south Sudanese running talent outside of southern Sudan.

The first I ever wondered about Sudanese athletes was when I wrote a story about Macharia Yuot, a runner at Widener University in Pennsylvania who caught my eye by winning the 2006 Division III cross-country championship in Wilmington, Ohio, before jumping on a plane that evening and finishing sixth in the Philadelphia Marathon—his first run longer than twenty-one miles—the very next morning. Yuot had been one of the "Lost Boys of Sudan," the largest contingent being from the Nilotic Dinka, who fled the violence that engulfed their homes. When he was nine, Yuot's town was overrun by the religious civil war that cost two million Sudanese their lives between 1983 and 2005. Rather than see their sons forced to walk minefields in order to clear the way for soldiers, parents bid them flee. So the boys walked the desert, alone. By 1991, some, like Yuot, who survived the soldiers hunting them—and the lions that occasionally carried away a sleeping boy—made it to a refugee camp in Kenya. In 2000, the U.S. government airlifted around 3,600 of the boys to America, and sprinkled them around the country with foster parents.

The Lost Boys had hardly unpacked by the time they started appearing in local newspaper headlines for their exploits on high school track teams. "Only months after settling in Michigan, two Sudanese refugees are finding that they are among the fastest high school runners in the state," went the lead of one AP article. Another, in the *Lansing State Journal*, noted that Abraham Mach, a Lost Boy who had no competitive running experience before arriving at East Lansing High, was the most outstanding performer in the thirteen-to-fourteen age group at the 2001 National AAU Junior Olympic Games, medaling in three events. Mach, who had been living in a Kenyan refugee camp just one year earlier, went on to become an NCAA All-American at Central Michigan in the 800-meters.

A cursory search of newspaper articles revealed twenty-two Sudanese Lost Boys mentioned for having run well in America in high school, college, or road races. The most prominent Lost Boy runner is Lopez Lomong, who in 2008 was a 1,500-meter runner and had the honor of bearing the U.S. flag at the Olympic opening ceremony in Beijing. In 2012, Lomong again made the U.S. Olympic team, this time in the 5K. In March 2013, he ran the fastest indoor 5K ever by an American citizen.

Not too shabby for a group the size of a large high school. And as soon as South Sudan became an independent country in 2011, it had an Olympic marathon qualifier in Guor Marial, who had fled Sudan for the United States and ran for Iowa State. Because South Sudan had not set up a national Olympic committee, and because Marial refused to represent Sudan, he was—following a hefty dose of public pressure on the International Olympic Committee—given special status and allowed to compete in London under the Olympic flag. South Sudan, then, doesn't even have an Olympic committee, but it has already had an Olympic marathoner.

All this is, of course, no more scientific than John Manners's time trial observations. In only slightly more scientific fashion, a few researchers and running enthusiasts have used statistics to suggest that the dominance of East African runners likely has a genetic basis. Anthropologist Vincent Sarich used world cross-country championship results to calculate that Kenyan runners outperformed all other nations by 1,700-fold. Sarich made a statistical projection that about 80 out of every 1 million Kenyan men have world-class running talent, compared with about 1 out of every 20 million men in the rest of the world. (The number would be far more staggering if he focused only on the Kalenjin.) A 1992 *Runner's World* article noted, based purely on population percentages, the statistical chances of Kenyan men having won the medals they did at the 1988 Olympics was 1 in 1,600,000,000.

Those are intriguing calculations, but without context do not shed much light on whether the natural gifts required for world-class running are more prevalent among Kenyans. German teams won the team dressage competition at every Olympics from 1984 to 2008, which, on a strictly population basis, is very unlikely. Still, we can all probably agree that German equestrians probably don't have dressage genes in greater frequency than is found among equestrians in neighboring European countries. But dressage is not a mass participation sport, so, frankly, any nation that is trying hard—German dressage was partly funded by the horse breeding industry—will do well. Canada produces the most NHL players because Canada invented ice hockey and, really, how many countries even have significant participation in hockey? The answer: not all that many. Or consider baseball's World Series, which is anything but a *world* series.

Plus, for years, the rest of the world was helping Kenya by getting slower. Even before Kenya commandeered the international running scene, the countries that had dominated distance running—Britain, Finland, the United States—were growing increasingly wealthy, increasingly overweight, increasingly interested in other sports, and increasingly less likely to train seriously in distance running. Between 1983 and 1998, the number of U.S. men who ran under 2:20 in the marathon for the year declined from 267 to 35. Great Britain declined from 137 to 17 over the same time period. The American nadir was 2000, when the United States qualified only one man for the Sydney Olympic marathon. Finland, which was the top distance running power in the world between World Wars I and II, when it was a poor rural country, did not qualify a single distance runner in any event at the 2000 Olympics. As Brother Colm O'Connell, a Patrician brother who came from Ireland to Kenya to teach high school in 1976 and stayed to coach elite runners—including current 800-meter world record holder David Rudisha—told me: "The genes didn't go away in Finland, the culture did."

A few countries held steady from the 1980s through the millennium, like Japan, which has between 100 and 130 sub-2:20 men just about every year. Meanwhile, Kenya jumped from a single sub-2:20 man in 1980 to 541 in 2006. (Kenyan marathoners really exploded in the mid-nineties, as the notion in Kenya that marathon training caused male infertility receded, and after Kenya's sports commissioner, KenSAP's own Dr. Mike Boit, allowed agents into the country and alleviated travel restrictions on athletes.)

Here's the conclusion of Peter Matthews, the track-and-field statistician who compiled those numbers: "In these days of computer games, sedentary pursuits, and driving our children to school—it is the 'hungry' fighter or the poor peasant who has the endurance background, and the incentive to work on it, who makes the top distance runner."

13

The World's Greatest Accidental (Altitudinous) Talent Sieve

S ugar, some sugar," he says. I must have looked confused. "You know, *sugar*. I appreciate."

We were standing, the runner and I, in Iten, Kenya, on a dirt track in Kamariny Stadium. But to call Kamariny a stadium is to elevate a sandlot to a cathedral. On one side is a wooden bleacher painted sky blue and crooked as rotten teeth. On the other side is a sheer cliff four thousand feet above the Rift Valley floor and eight thousand feet above sea level. Dozens of runners circle the track in interval sessions, as a sheep wanders over the escarpment to graze on the infield.

The runner I'm speaking with is twenty-four-year-old Evans Kiplagat, and he wants me to buy him sugar. Earlier that Thursday morning, Kiplagat loped six miles to the track, then ran a hard workout. In a few minutes, he'll embark on the six-mile trip home. If I don't buy him food, he'll return hungry to the wooden shed that a local man lets him use on his *shamba*—a subsistence farming plot.

Kiplagat's parents did not own the *shamba* on which they lived, so when they both died of illnesses in 2001 he couldn't remain on the land. He is thankful for his current room, "but food is a problem," he says. Most Tuesdays and Thursdays, Kiplagat jogs to the track and latches on to a training group that includes an athlete like Geoffrey Mutai, a Boston and New York City marathon champion, or Saif Saaeed

Shaheen (formerly Stephen Cherono), the world record holder in the 3,000-meter steeplechase who was raised and trains in Kenya but was paid to switch his citizenship to Qatar. After the workout, Kiplagat will log more miles, walking among his friends' homes to see if anyone has leftover *ugali*, the doughy cornmeal that rural Kenyans eat every day of the week. If he scrounges up enough food, he'll go for another six miles in the evening. For Kiplagat, every day is a two-a-day, if not a three-a-day, and that doesn't count running six miles each way as transportation to and from the track on Tuesdays and Thursdays.

It is the schedule of a man who burns to run, whose passion is to compete on the highest stage, and to stand atop the podium and weep at his national anthem. Except, that's not who Kiplagat is.

"If you could get a job in the military, would you stop training?" I ask.

"Yes."

"What about a job with the police?"

"Yes. Any job," he says.

Kiplagat would prefer to have a job that would allow him to continue training, but he is content to stop running *tomorrow* if someone offers him a decent living. He started training in 2007 after walloping his high school friends in a small race. Last year, Kiplagat ran 29:30 in a hilly 10K road race in Kenya, an outstanding time relative to most of the world, but not one that makes him stand out from the Kamariny Stadium crowd. So he'll keep trying to borrow enough money to travel to Kenyan cities for races so that he can post a time that will attract an agent.

There are Evans Kiplagats all over Kamariny—about one hundred runners were training at the track the day I was there—striding right alongside world champions. Sporadically, an unfamiliar man will wander onto the track and right away try to keep pace with Olympians. If he holds up, perhaps he'll come back. If not, he'll slink back to the *shamba*. It's a microcosm of the overall training scene in Kenya: there are few training secrets here—some top runners don't even have coaches—but there are hordes of runners willing to train multiple times a day, as full-time athletes. In the United States, a top college distance runner usually has to put off making a living for a few years in order to chase a dream. "In Kenya, it's just the opposite," says Ibrahim Kinuthia, a former international runner and now a coach in Kenya. There is no career or grad school to delay, and thus no opportunity cost for most rural Kenyans to take a crack at training with the elites.* Given Kenya's annual per capita income of $800, according to the World Bank, the potential payoff for running success is greater, relatively speaking, than even an NBA contract is for an inner-city American boy. Winning a single major marathon brings a six-figure payday. Even earning a few thousand dollars in smaller road races in America and Europe is a relative windfall for most rural Kenyans. Successful runners quickly become one-man or one-woman economies. In Eldoret, the major city near Iten and Kamariny Stadium, Moses Kiptanui, the former steeplechase world record holder, owns a dairy farming business. He also owns the trucks that transport the milk, and the building in town with the supermarket that sells the milk. The result of these economic incentives is an army of aspiring runners who undertake training plans fit for Olympians, with many falling by the wayside, and those who survive becoming professionals.

Interestingly, a system that thrives on the hard work of many is fueled by an abiding belief in natural talent. The Kenyan coaches and runners I spoke with almost uniformly said that it was never too late to begin training. If one has talent, they said, then one just needs to start training hard and elite status will come swiftly.

A number of Kenya's most luminous running stars have succeeded precisely because they did not assume it was too late. In a hotel in Nairobi I met Paul Tergat, former marathon world record holder and

*Until recently, married Kenyan women were essentially barred from training. But, as Kenyan women have won major paydays on the international circuit, "there [has been] a total change in the way to imagine the possibility of a woman training in Kenya," says Gabriele Nicola, an Italian who coaches top Kenyan women. "Before, in Africa, the idea was the girls are weaker than the men." But that is changing rapidly. Nicola thinks it will take about another decade before Kenyan women have completely banished the perception that they are unfit for rigorous training.

the greatest cross-country runner in history, who told me that he played volleyball in high school and didn't start running until "between nineteen and twenty, when I started the military. There I met a number of great runners I used to read about, like Moses Tanui and Richard Chelimo. So I trained, and by twenty-one I realized I had the talent." And by twenty-five he had won the first of five consecutive world cross-country championships.

The similarity to Jamaican sprinting—or to Canadian hockey, or to Brazilian soccer—is that there is a large number of athletes put in the top of the funnel, and a smaller number who display talent and survive the rigorous training and come out the bottom as world beaters.

While some of Kenya's best runners have entered the game very late, for scores of others, training starts very early in life, before they even know it.

Kenya is particularly harrowing for Yannis Pitsiladis, the University of Glasgow biologist whose passion is collecting the DNA of elite athletes. Given his fear of flying, he drives all over Kenya. Navigating the pocked roads of rural Kenya is like guiding a marble through the game Labyrinth. (Eventually, you're going to lose.) And yet, for a decade, Pitsiladis has returned to Kenya over and over. As expected from the Kalenjin running hotbed, he and his colleagues have found that individuals with genes that indicate Nilotic ancestry are vastly overrepresented among elite athletes. But, as in Jamaica, the findings that have most affected him are cultural, not genetic.

Pitsiladis's work has shown that international-level runners from Kenya are most often of the Kalenjin tribe, most often from poor, rural areas, and very likely to have had to run to school growing up. In one study Pitsiladis conducted with colleagues, 81 percent of 404 Kenyan professional runners had to run or walk a considerable distance to and from primary school as children. Kenyan kids who rely on their feet to get to and from school have 30 percent higher aerobic capacities on average than their peers. World-class athletes were also more likely than lesser athletes to have had to run or walk six miles or more to school. Pitsiladis talks fondly of one ten-year-old boy who was such a proficient runner that he took off at a six-minute-mile pace during a test of his aerobic capacity on a dirt track.

When I visited Kenya, running up and down the red dust hills of Iten, the epicenter of Kalenjin training, occasionally kids would join by my side, excitedly chirping their favorite English phrase: "How are you! How are you!" On my last run in Iten, a boy who looked about five years old tagged along as I trudged up a long hill. The boy was in ragged sandals and carrying a loaf of bread under one arm. He followed for a few minutes and then slunk under a wooden fence, tugging his loaf of bread behind him, and disappeared. It struck me that there is no such thing as a casual jogger in Kenya, only those who run for transportation, those who are killing themselves in training, and those who are not running at all.

After that run, I mentioned the bread boy to Harun Ngatia, a physiotherapist who treats Kenyan pros. "When the boy grows up," he said, "all he will know is running." His words reminded me of a late-1990s mock charity drive declared on a now-defunct online track-and-field message board: Help Americans compete in distance running by donating school buses to Kenyan children.

And it isn't just Kenya. Pitsiladis and a research team found a similar pattern in the world's second distance running superpower, Ethiopia. As in Kenya, Ethiopian runners tend to come from a traditionally pastoralist ethnic group—the Oromo—and they are also much more likely to have had to run to school than nonrunners, and professional Ethiopian marathon runners are more likely to have had to run long distances to school than professional Ethiopian 5K and 10K runners. Meanwhile, analysis of the mitochondrial DNA of Ethiopian and Kenyan runners shows that their maternal lines are not especially closely related. So there is no single, genetic supertribe of runners that is contiguous from Ethiopia to Kenya. (Ethiopians tend to have more chunks of mitochondrial DNA found in Europeans, possibly reflective of Ethiopia as the original point of migration for all humans outside of Africa.)

No one has conducted a study of the running economy of untrained Ethiopian children, as the Danish scientists did in Kenya, so it is unknown how the Oromo compare with the Kalenjin in that respect, but it is clear that these two groups both embrace running as a way of life. "You have all these kids running," says Pitsiladis, "and then a boy or girl sees that they can run faster than the others. You absolutely must have the right genes. You must choose your parents correctly, but you have thousands of kids running and the cream rises to the top. After ten years of work, I have to say that this is a socioeconomic phenomenon."

When I asked Ethiopian icon Derartu Tulu—Olympic 10K gold medalist in 1992 and 2000—if any of her two biological or four adopted kids like to run with her, she replied: "No, they say they get tired when I take them training with me. They don't like to run. . . . I think it is because they go to school by car." Says Moses Kiptanui, the Kenyan former steeplechase world record holder, of his children: "A vehicle came and took them to school . . . they like to do easier sports."

"How many of the top Kenyan runners have sons or daughters who are excelling at running?" Pitsiladis asks, rhetorically, after noting that there are plenty of Kenyan siblings and cousins who excel. "Almost none. Why? Because their father or mother becomes a world champion, has incredible resources, and the child never has to run to school again."

Still, it would be an unfair stereotype to suggest that all great Kenyan athletes ran to school, as there are conspicuous exceptions, like Paul Tergat, the greatest cross country runner in history. "I think the majority of us are running to school barefoot," Tergat says. "But my school was very close. I could walk to school." And the same goes for Wilson Kipketer, one of the greatest middle-distance runners of all time, whose school was next door to his home. Both men were world record holders, so, clearly, running to school is not a necessary trait of a world record holder. Nor is it sufficient. A few of the Kenyan children that Pitsiladis has tested who run miles to school nonetheless have pedestrian aerobic capacities, reminiscent of the low responders in the HERITAGE Family Study. "It's a small number," he says, "but there are some." Not to mention that millions of Kenyan children across the country travel to school on foot, and yet the Kalenjin still stand apart in their running success.

Pitsiladis believes adamantly that in addition to the large number of running kids, there is another essential component to Kenyan running success. It is exactly what the Rift Valley ledges that are home to both the Kalenjin in Kenya and the Oromo in Ethiopia share: altitude. "You must live at altitude," Pitsiladis says. "Some have said that the best way is to live high and train low. The Kenyans live high and train higher."

"If it's just the altitude, where are the runners from Nepal?" Brother Colm O'Connell asked, while sitting in his home in Iten, as 800-meter world record holder David Rudisha sank into the couch.* In the backyard is "the gym," a single metal pole dipped in cement at both ends so it resembles a barbell.

At the very least, the altitude along the Rift Valley rim—where mosquitoes are scarce—likely prevented Kenyan runners who live there from the distance running *disadvantage* of genetically lowered hemoglobin, which occurs in people with ancestry in malaria danger zones.

But O'Connell's question is intriguing, and has been asked rhetorically for years about the Kenyan running phenomenon. Altitude is known to increase red blood cells in athletes who move from sea level to the mountains, so why, then, aren't runners coming down from the Andes and the Himalayas and smoking the rest of the world, as the Ethiopians and Kenyans have done?

The "Nepali runners" question, though, is actually irrelevant to the Kenyan and Ethiopian running phenomena, and not only because the Himalayan climate does not foster a narrow body type. One clear point

---

*Rudisha is a member of the Masai ethnic group. (His mother, though, is Kalenjin, and his Olympic medalist father is part Masai.) The Masai are also a Nilotic people and relatively closely related to the Kalenjin. According to data in Jean Hiernaux's *The People of Africa*, the Masai have extremely long legs in proportion to their height.

of science is that the genetic means by which people in different altitudinous regions of the world have adapted to life at low oxygen are completely distinct. In each of the planet's three major civilizations that have resided at high altitude for thousands of years, the same problem of survival is met with different biological solutions.

By the late nineteenth century, scientists figured they understood altitude adaptation. They had studied native Bolivians, living in the Andes at higher than thirteen thousand feet. At that altitude, there are only around 60 percent as many oxygen molecules in each breath of air as at sea level. In order to compensate for the scarce oxygen, Andeans have profuse portions of red blood cells and, within them, oxygen-carrying hemoglobin.

The amount of oxygen in the blood is determined by two factors: how much hemoglobin one has and its "oxygen saturation," or how much oxygen that hemoglobin is carrying. Because there is so little oxygen in their air, many of the hemoglobin molecules in the blood of the Andean highlanders rush through the body without a full load of oxygen—like roller coaster cars with few passengers. But the Andeans make up for it by having many more cars. This is not necessarily good from an athletic standpoint. Andeans have so much hemoglobin that their blood can become viscous and unable to circulate well, and some Andeans develop chronic mountain sickness.

Nineteenth-century scientists also saw that Europeans who traveled from sea level to altitude responded the same way, by producing more hemoglobin. So the book on altitude adaptation was closed for almost a century—until the 1970s, when Nepal and Tibet began to open to foreigners.

Cynthia Beall, an anthropology professor at Case Western Reserve University in Cleveland, started visiting to study Tibetans and Nepalese Sherpas who can live as high as eighteen thousand feet. To her surprise, Beall found that Tibetans had normal, sea-level hemoglobin values, and low oxygen saturation, lower than people at sea level. *Few roller coaster cars, and many of them weren't full.*

Most Tibetans have a special version of a gene, EPAS1, that acts as a gauge, sensing the available oxygen and regulating the production of red blood cells so that the blood does not become dangerously thick. But it also means Tibetans don't have the increase in oxygen-carrying hemoglobin that Andeans do. "So, how exactly are they surviving here?" Beall asked herself. "The oxygen in their blood seems very low, but they're somehow delivering enough to function normally."

Eventually, Beall determined that Tibetans survive by having extremely high levels of nitric oxide in their blood. Nitric oxide cues blood vessels in the lungs to relax and widen for blood flow. "The Tibetans have 240 times as much nitric oxide in the blood as we do," Beall says. "That's more than in people at sea level who have sepsis," a life-threatening medical condition. So Tibetans adapted by having very high blood flow in their lungs, and they also breathe deeper and faster than native lowlanders, as if they're in a constant state of hyperventilation. "They're spending more energy doing that," Beall says.

In 1995, Beall and a team moved on to the remaining population in the world that has lived at high altitude for thousands of years: Ethiopians, and specifically the Amhara ethnic group living at 11,600 feet along the Rift Valley. Yet again, she found an altitude biology unique in the world. The Amhara people had normal, sea-level allotments of hemoglobin and normal, sea-level oxygen saturation. *The same number of roller coaster cars as sea-level natives and nearly all of them were filled, just as in sea-level natives.* "If we didn't know we were at altitude, I would've said we were looking at sea-level people," Beall says. It's not entirely clear how the Amhara pull this trick off. But Beall has preliminary data on Amhara Ethiopians that shows they move oxygen unusually rapidly from the tiny air sacs in their lungs into their blood.

New Zealand's Peter Snell, former mile world record holder turned medical researcher, theorized that enhanced transfer of oxygen from the lungs to the blood might be an advantage for people with altitude ancestry when they came to run at sea level. "It's possible," Beall says of that prospect. She once raised it in a paper, but she's adamant that nobody really knows. Plus, she saw the enhanced oxygen diffusion in her Amhara data, and most of the top Ethiopian runners are Oromo. An Oromo man holds the world records in the 5K and 10K, and an Oromo woman holds the women's 5K record. (Scientists tracked the Oromo man, Kenenisa Bekele, over two runs at 6:30-per-mile pace, one at just under five thousand feet, and one above ten thousand feet. Astoundingly, his average heart rate only increased from 139 beats per minute to 141 on the higher run.)

Unlike the Amhara, who have been at altitude for thousands of years, Beall says that the pastoralist Oromo moved up from sea level just five hundred years ago. A foreigner would not distinguish Amhara and Oromo people on sight, but in terms of their altitude response, Beall would never confuse them.

Beall tested Oromo people living at about the altitude of Denver, "so you wouldn't expect to see much," she says, in terms of elevated hemoglobin. "But they already had more than a gram of hemoglobin more than the Amhara at a comparable altitude." And the hemoglobin was packed with oxygen. "Their hemoglobin level was definitely higher than you would expect from a random group of lowlanders," she says. Whereas the Amhara had low hemoglobin even at high altitudes, the Oromo had high hemoglobin even at moderate altitudes.

For one, these differences emphasize the diversity of physiology between peoples who have lived at altitude for different spans of history, and for whom evolution has landed on novel genetic solutions. Himalayans and Amhara Ethiopians are thought to have lived at altitude for thousands and perhaps tens of thousands of years, and Andeans for a shorter span, which may explain why Andeans are not yet fully adapted to their extraordinarily high homeland—and why they greatly elevate hemoglobin, just as lowlanders who go to altitude. (Like the Oromo, Kenya's Kalenjin are relatively new altitude dwellers, having settled at altitude no more than two thousand years ago.)

As for Beall's data from the Oromo—the ethnic group of the majority of top Ethiopian runners—they smack of altitude *responders*. The Oromo she tested increased their hemoglobin markedly even at altitudes below a mile high. And not only do different ethnic groups respond biologically to altitude in unique ways, there is also tremendous variation among individuals from the same ethnic group.

In 2003, a team of scientists from Norway and Texas exposed athletes to 9,200 feet of altitude for one day and looked at changes in the levels of the hormone EPO—which spurs the body to produce red blood cells. (Cheating endurance athletes inject EPO in an effort to force their bodies to produce more red blood cells.) The variation ranged from an athlete whose EPO levels declined, to another whose levels increased more than 400 percent.

In separate work on runners who trained for a month at altitude, those whose supply of red blood cells increased 8 percent on average improved their 5K time by thirty-seven seconds upon returning to sea level, whereas those who had no increase in red blood cells did slightly worse than they had previously in the 5K when they returned to sea level. As with other forms of training—and all manner of medicine—altitude training is most effective if tailored to each athlete's unique physiology.

The idea of individualized responses to altitude rings true to Bob Larsen, who coached Americans Deena Kastor and Meb Keflezighi—winners, respectively, of a bronze and a silver medal in the 2004 Olympic marathon. "We have some evidence that some people have to be there for a long time," Larsen says. "It really took Deena about two years of being at altitude. Meb was quick. He was a little flat his second week at altitude, but after about six weeks he set the American record [in the 10K]."

Even with individual variation in altitude response, there seems to be a rough "sweet spot" for training, an altitude where red blood cell production increases, but not too much. Where the air is thin, but not too thin. Andeans and Himalayans live far above it. Anecdotally, the sweet spot is around six to nine thousand feet, high enough to cause physiological changes, but not so high that the air is too thin for hard training.

As it happens, the ridges of the Rift Valley in Ethiopia and Kenya are plumb in the sweet spot. The foremost training bases in Kenya: Eldoret, 6,890 feet. Iten: 7,545 feet. Kapsabet: 6,395 feet. Kaptagat: 7,870 feet. Nyahururu: 7,215 feet. The major training cities in Ethiopia, Addis Ababa and Bekoji, both have running sites around 8,000 to 9,000 feet. In the United States, pro endurance athletes hunting for the sweet spot train in Mammoth Lakes, California: 7,880 feet. Or Flagstaff, Arizona: 7,000 feet.

Preferable to moving to altitude to train is being born there. Altitude natives who are born and go through childhood at elevation tend to have proportionally larger lungs than sea-level natives, and large lungs have large surface areas that permit more oxygen to pass from the lungs into the blood. This cannot be the result of altitude ancestry that has altered genes over generations, because it occurs not only in natives of the Himalayas, but also among American children who do not have altitude ancestry but who grow up high in the Rockies. Once childhood is gone, though, so too is the chance for this adaptation. It is not genetic, but neither is it alterable after adolescence.

No scientist contends that altitude alone forges tireless runners or that it is impossible to become a great distance runner without altitude training. But some, like Pitsiladis, say that it's simply far less likely. A helpful combination, perhaps, is to have sea-level ancestry—so that hemoglobin can elevate quickly upon training at altitude—but to be born at altitude, in order to develop larger lung surface area, and then to live and train in the sweet spot. This is exactly the story of legions of Kalenjin Kenyans and Oromo Ethiopians.

Coincidentally, or maybe not, Shalane Flanagan, the fastest current American female marathoner—and daughter of a former marathon world record holder—was born and spent part of her childhood in the foothills of the Rockies, in Boulder, Colorado, above a mile high. Ryan Hall, the fastest current male American marathon runner, was raised in Big Bear Lake, California: seven thousand feet, and up.

Drive north toward the Sangre de Cristo Mountains, to the point where the black asphalt disappears beneath a wash of brown rock and dirt, and you will be in Truchas, New Mexico, at eight thousand feet.

Not long before the road vanishes, on the left just past a cattle gate, is a low-slung adobe house with a yellow school bus in the yard. The bus hasn't moved in decades. In the alfalfa field out back, eighty-five-year-old Presiliano Sandoval is working in the heat. His fingers, which haven't been parallel since before the school bus worked, are curled around the wooden handle of a shovel.

In the adobe house, Presiliano raised the greatest American athlete no one remembers. Even now, Anthony Sandoval lives just an hour's drive to the southwest, in Los Alamos. Anthony was one of six children, but Presiliano could tell that he was different. Presiliano remembers Anthony, at eight years old, was content to walk alone in the winter into the mountains with a hammer and wedge to split frost-hardened piñon trees.

By the summer of sixth grade, three times a week Anthony was taking his father's cows several miles into the mountains so they could graze. "It was never less than two hours of walking," with sporadic running mixed in, Anthony says. He had always been a good runner, but when he returned from that summer, he was by far the fastest boy in school.

Presiliano yearned for his son to get an education that Truchas could not provide, so he enrolled him at Los Alamos High School an hour away, where Anthony was surrounded by the sons and daughters of the physicists and nuclear engineers who worked at Los Alamos National Lab, birthplace of the atomic bomb. The locale was so secretive during World War II that babies born in Los Alamos had "P.O. Box 1663" listed as the city of record on their birth certificates.

At the start of Sandoval's freshman year, a friend suggested he go out for cross country. "I said, 'What's cross country?'" Sandoval recalls. "But I went out that year, and ended up second in the state. And then I never lost another race after that in high school." In his junior year, Sandoval ran farther than 12.5 miles in 60 minutes, setting the under-twenty world record for a one-hour run. In 1972, his senior year, then 5'6" and 98 pounds, Sandoval won the junior national championships in cross-country.

The Sandovals had no phone in the adobe home in Truchas, but reams of recruiting letters were mailed straight to Los Alamos High School. The boy whose aunts and uncles had been shepherds and uranium miners would go to Stanford. In Palo Alto, Sandoval excelled in class, earning admission to med school while training sixty to seventy miles a week.

At the Pac-8 championships in 1976, his senior year of college, Sandoval won the 10K, just ahead of three Kenyans running for Washington State, one of whom would later set the world record. And then, off his college track training, Sandoval jumped into the 1976 Olympic marathon trials. He finished fourth, one minute and one spot off the Olympic team. So away he went to med school, figuring he would have other chances at the Olympics, when he could actually train for the marathon distance.

But Sandoval was insatiably interested in serving people and in medicine, so he pursued cardiology, a study-intensive specialty that did not accommodate marathon training. Still, Sandoval's ability was evident. In 1979, immersed in his medical studies, Sandoval managed just thirty-five miles per week of training. It was enough for him to run a 2:14 marathon, an utterly preposterous result given what was essentially a jogger's training regimen. (A baseball fan might think of this as akin to a guy who takes batting practice in his local beer league and then hits .300 against major league pitchers.)

In 1980, with the Olympics again approaching and still deep in med school, Sandoval carved out a few months of rigorous training. It was enough. At mile 23 of the Olympic Trials in Buffalo, he simply ran away. He finished in 2:10:19, a U.S. Olympic Trials record that stood for twenty-seven years. "Tony was, at that point, probably the fastest runner in the world," says Frank Shorter, the last American man to win gold in the Olympic marathon.

But it was the year of the Moscow Olympic Games, and President Jimmy Carter decreed that the U.S. would lead sixty-four countries in an Olympic boycott to protest the Soviet invasion of Afghanistan. Sandoval, like 465 other American athletes, was forced to stay home.

As he embarked on his career as a cardiologist, Sandoval began a pattern that would last more than a decade: he would try, but struggle, to ramp up his training each time the Olympics approached. In 1984, he finished sixth at the trials. In 1988, in the middle of a cardiology fellowship, he finished twenty-seventh.

As the 1992 Olympic Trials approached, Sandoval, by then thirty-seven, realized it would be his last shot. Finally, he took time off to train, and was in phenomenal shape. On a warm, windy day in Columbus, Ohio, he felt effortless through the opening miles. "I was in heaven," Sandoval says. "I was thinking, 'This is my fifth Olympic Trials, and this is going to be a good day.'" And it was, until he planted his foot to make a turn at the bottom of a hill around mile eight and felt pain shoot down the back of his leg. "I figured it was my calf, so I stopped to massage it," Sandoval says. "I was watching the time. I was in such good shape, I figured I could give the leaders about two minutes and still make the team." By mile 13, his leg was swelling and he could hardly walk. He hobbled off the course. "I knew it was over then," Sandoval says, quietly, "that I'd never go to the Olympics." He had run five miles on a ruptured Achilles tendon.

Today, in an office across the street from his high school track, Sandoval is one of just a few cardiologists serving all of rural northern New Mexico. In his home, Sandoval still has the blue velour USA outfit he would have worn at the 1980 Olympics. "It just hurts when you start thinking about it," Sandoval says. "I never got to run as hard as I can." His voice halts when he mentions how proud his six children—all college athletes—would have been to see Dad's medal. "I think sometimes he wishes he'd taken more time off from medicine to train," says Sandoval's wife, Mary.

Even now, Sandoval is thin enough to hide behind a parking meter, and by 6:30 A.M. most mornings he is skimming along forest switchbacks in the nearby Jemez Mountains. There is no wasted movement in his stride. His arms are carried high and tight. He seems barely to come off the ground, sweeping over the soil as lightly as a water bug flitting across a pond. He refers to some of the trees and rock outcrops along the trails as "old friends."

David Martin, former head of USA Track and Field's physiological testing program, studied Sandoval back when he was competing. "Anthony was a physiological specimen to be reckoned with," Martin says. "He had long legs, a huge heart, huge lungs, and a small torso. I tested him at my lab in Atlanta, and boy could he move oxygen. I don't want to say Anthony is a genetic freak, but he's unusual, because even as he got older his body size remained diminutive and his heart size increased."

Martin pauses, and considers Sandoval in his entirety. His quiet toughness. His lithe body. His huge aerobic capacity. His rural youth at eight thousand feet, and his childhood of running and walking for transportation. He clearly had physiological gifts, but he also had a unique crucible in which to discover and develop them.

"You know what he is?" Martin asks, awaking excitedly from a pensive moment. "He's a Kenyan, that's what he is! He's an American Kenyan."

Eldoret is a bustling city of 250,000 near the heart of the Kalenjin training region in Kenya. The occasional donkey cart jockeys with cars for right of way as they navigate the rutted roads. The rush in the street is frantic. Shoppers hustle in and out of ground-level stores or the eateries above them. Narrow alleys are stuffed with hole-in-the-wall shops. Here you can buy Nike running shoes that were brand-new fifteen years ago but are still unworn, because Kenyan professional runners will sell the shoes they receive from sponsor companies to resellers as soon as they get them. In one alcove, a man furtively peddles Kenyan national team gear out of a backpack.

One day while I was in Eldoret, I sat in a garden behind a steel guard wall and had Kenyan tea—which has milk and sugar—with Claudio Berardelli, a young Italian who moved to Kenya and has become one of the world's top coaches of distance runners. Berardelli was coauthor of a paper that was about to come out in the *European Journal of Applied Physiology*. The paper looked explicitly at running economy, comparing 2:08 European marathoners with 2:08 Kalenjin marathoners. Not surprisingly, the physiologies of the runners—their aerobic capacities and running economies—were very similar. The authors concluded, then, that superior running economy does not explain the dominance of Kalenjin marathoners over Europeans.

In reality, though, they did not ask a question that could provide such an answer. It is no surprise that 2:08 marathoners look physiologically similar no matter their nationality or ancestry. After all, they are all 2:08 marathoners. The question is whether there are many more people in one place who are capable of becoming 2:08 marathoners than in another or why 2:03 and 2:04 marathoners come only from Kenya and Ethiopia.

Berardelli's opinion, outside of the paper, was very different from the conclusion in it: "I don't believe that in Italy there is not somewhere another [Stefano] Baldini," he says, referencing the Italian who won gold in the 2004 Olympic marathon. "And Italians are probably saying, 'There is no need to look for him, because Kenyans always win.' So they don't find him." But does he think there are as many potential Baldinis in Italy as in Kenya? "I think in Kenya maybe you will find ten Baldinis, and in Italy maybe you find two Baldinis. But *come on guys,* work on finding them!" So, then, Berardelli's opinion is that gold-medal marathon potential is not exclusive to Kenya, but that it is more common there. "I think the Kenyan lifestyle probably fixed genetically some characteristics good for running," he says.

And while a naturally narrow body type is crucial to running economy, economy can also be improved. There is no better example than the greatest female marathoner of all time, Britain's Paula Radcliffe. Radcliffe entered her first races at nine, though she hadn't begun real training. By seventeen, Radcliffe was a promising junior athlete, and Andrew M. Jones, a British physiologist, started working with her. Immediately, Jones saw that Radcliffe was gifted. There were outstanding athletes in her family—her great aunt Charlotte was an Olympic silver medalist swimmer—and she had a VO_2max essentially as high as elite female athletes ever get, even though she was training less than thirty miles a week. "Clearly, she was exceptionally talented," Jones wrote of Radcliffe. "However, this athletic potential was only achieved following ten further years of increasingly arduous training."

Over those years, Radcliffe got taller, but stayed the same weight as she trained maniacally, often at altitude. Her VO_2max did not improve at all—it was already at the top—but each year her running economy got incrementally better, presumably at least partly because her legs got longer while her weight did not increase. In 2003, eleven years after she was first tested, Radcliffe's VO_2max was no different from what it was when she was eighteen and training lightly, but her running economy had improved dramatically, and she shattered the women's world marathon record in 2:15:25. Obviously, Radcliffe's exceptional running economy was at least partly created by her training.[*]

Genetic science, even as it matures, is unlikely to provide anything resembling a complete answer to the questions behind Kenyan running prowess. Just as it is tough to find genes for height—even though we know they exist—it is extraordinarily difficult to pin down genes

---

[*]An additional hypothesis, mentioned by physiologists I spoke with, is that Radcliffe's Achilles tendon had hardened over years of training—like high jumper Stefan Holm's—improving her running economy.

for even one physiological factor involved in running, let alone all of them. As Sir Roger Bannister, a world-renowned neurologist and the first man to break four minutes in the mile, once said: "The human body is centuries in advance of the physiologist, and can perform an integration of heart, lungs, and muscles which is too complex for the scientist to analyze."

Additionally, the prevalence of gene variants differs to such an extent among ethnicities that geneticists use ethnically matched control subjects in their studies. So a Kalenjin genetic study uses Kalenjin runners and compares them with Kalenjin controls. Thus, genetic studies usually look for differences *among* members of an ethnic group, and usually say little about differences *between* ethnic groups. With the physiology of running far from fully understood, we should not hold our breath for genetic technology in itself to solve the Kenyan question, at least not anytime soon. We will have to look other places for insight, as did the Danish researchers who tested running economy in Kalenjin boys.

When I last spoke with Berardelli he had just begun to coach a group of Indian athletes who came to Kenya to train. On the face of it, they have incredible environmental similarities to his Kenyan runners: impoverished backgrounds, high motivation, and childhoods filled with running for transportation. If distance running success requires only monetary incentives, childhood running, and world-class training, then expect to see some of Berardelli's Indian pupils alongside the Kenyans soon.

"So," Berardelli said, with a doubtful smirk. "We will see."

Berardelli believes that the Kenyans are, in general, more likely to be gifted runners. But he also knows that no matter their talent or body type or childhood environment or country of origin, 2:05 marathoners do not fall from the sky. Their gifts must be coupled with herculean will.

Although that, too, is not entirely separable from innate talent.

14

Sled Dogs, Ultrarunners, and Couch Potato Genes

The aluminum sign for Comeback Kennel is nailed haphazardly to an evergreen tree north of Fairbanks, Alaska, off the Elliott Highway and two miles in on a dirt road. The gravel driveway is packed hard from the cold and steep enough to make entry without an SUV precarious. This solitary spot befits Alaskan taste. If you can see the smoke from your neighbor's chimney, he probably lives too close.

It's an unlikely address for a collection of the planet's greatest and most steel-willed endurance athletes. But there, on a sloped clearing framed by black spruce, are 120 of dogsled racing's most distinguished Alaskan huskies. Comeback Kennel is really just the name of the frosted front yard belonging to Lance Mackey.

Mackey is an icon in the dogsled racing world, where he essentially invented the thousand-mile double. That is, in both 2007 and '08, Mackey won the thousand-mile Yukon Quest, and then, just weeks later, the world's other thousand-miler: the Iditarod, known to the faithful as "The Last Great Race on Earth." Prior to Mackey's back-to-back doubles, the feat was thought to be impossible. A musher was lucky to escape even one of the races without illness or serious injury to himself or his dogs. Even if he did, there is the problem of will, for both the dogs and their master.

Eminent mushers have had to withdraw from the Iditarod when

their dogs simply lie down in the snow and refuse to go another step. And the freezing cold and sleep deprivation of the long Alaska nights are famous for divorcing Iditarod mushers from their better judgment. From time to time, a musher crossing atop the frozen Bering Sea will gaze into the bright sunlight after a deep black night and start removing his jacket and gloves, only to be greeted by -50 degree air, and instant frostbite. Mackey himself has heard voices. Once, after a long, cold stretch with no sleep he was pleased to see an Inuit woman beside the trail smiling at him. He turned and started waving, and only then realized that she was gone. Or, rather, that she had never been there at all.

Prior to Mackey's runs, just to attempt to finish the Yukon Quest and the Iditarod back-to-back was considered foolhardy. Even if the musher survived the Quest with his vital signs intact, what about the dogs? Assuming they were healthy, would they want to keep running? Sled dogs, like their masters, must have the will to forge ahead.

"These aren't house dogs. Food will not work as a training device for sled dogs," says Eric Morris, a musher and biochemist who created Redpaw dog food for canine athletes. "Negative reinforcement will not work as a training device for sled dogs either. To go that distance, it's like a bird dog sniffing down a pheasant, it has to be the one thing in their life that brings them the greatest amount of pleasure. They have to have the innate desire to pull [the sled] . . . and you will find varying degrees of that in different dogs."

Each of the Alaskan huskies in Mackey's yard is chained to a metal ring that is looped around a pole, restricting its movement to a circle several meters in diameter that includes entrance into its own wooden house. Each dog, that is, except for Zorro.

On top of the hill in the yard is Zorro's fenced-in pen. He has more space, and no chain. It's his "condo on the hill," Mackey jokes. From here, Zorro looks down on the nighttime lights of Fairbanks far below, and also on his nieces, nephews, sisters, brothers, and sons and daughters, all here in this yard.

As Mackey walks toward Zorro, he pauses to point. "That's my main bitch right there," Mackey says, gesturing to one of Zorro's granddaughters, a female dog named Maple whose golden brown coat is the color of cinnamon toast. In 2010, Maple led Mackey's team—meaning she was at the head of the group of dogs—and won the Golden Harness Award for the most outstanding performer in the Iditarod. Like Maple, all of Mackey's champion dogs are in Zorro's line. "It was pretty ballsy," Mackey says, "to base the whole kennel around one dog." He leans down to nuzzle the blond rings of fur around Zorro's eyes, the ones that resemble the mask his namesake wore.

After communing with Zorro, Mackey walks back to the half-constructed house that he and his wife Tonya share. It's full of exposed wiring, and still partially wrapped in Tyvek sheets, but it belongs to them, along with the garage that holds a limited-edition Dodge Charger and three Dodge trucks, all prizes for Iditarod wins. "The dogs bought all of this," Mackey says. None more so than Zorro.

Zorro is the genetic nexus of the kennel, and not because he was a particularly fast husky. (He wasn't.) Rather, Mackey bred for the genes of work ethic. He had no other choice. In 1999, when Mackey began his breeding program, he couldn't afford the fastest, sleekest dogs.

Lance Mackey's father, Dick, was one of the cofounders of the Iditarod Trail Sled Dog Race, first run in 1973. In his first five attempts at the race, Dick never finished higher than sixth. In his sixth attempt, in 1978, something so unexpected occurred that the fledgling race had no rule to adjudicate it.

Lance was seven years old, standing near the burled arch that marks the finish, when his father, running alongside his sled and almost suffocating in his parka, sprinted down Front Street in a dead heat with defending champion Rick Swenson, also on foot beside his sled. As

Dick Mackey's lead dog crossed the finish first, by a nose, Mackey collapsed to the ground, leaving his team straddling the line as Swenson's sled zoomed past. At the end of 14 days, 18 hours, 52 minutes, and 24 seconds, the Iditarod had come down to whether race marshal Myron Gavin would rule that the first musher with a single dog across the line won or whether it was the musher with all his dogs across the line. "They don't take a picture of the horse's ass, do they?" Gavin asked, rhetorically. And so Dick Mackey won the Iditarod, and became a full-fledged hero to his son.

"I was standing right at the finish line," says Lance, who grew up in Wasilla, Alaska. "It was exciting. It was dramatic. It was emotional. It was embedded in my head. I have no doubt that something in that moment, in that one second, affected my passion or my drive or my commitment. It not only changed my dad's life, it changed mine." From that moment on, Lance Mackey always told himself that one day he would win the Iditarod too. But the path would be tortuous.

Three years after his father won the Iditarod, Mackey's parents divorced. He began to see little of his father, an ironworker who was off building up the remote reaches of Alaska. His mother, Kathie, worked as a bush pilot and dishwasher to support the family, so Lance had all the unsupervised time in the world to seek out trouble. He excelled at finding it.

By fifteen, Mackey was a one-boy crime wave: fighting, consumption of alcohol by a minor, more fighting, drunk and disorderly, public urination, and a little more fighting. Before he had a driver's license, he stole Kathie's checkbook, used it to buy a '68 Dodge Charger and drove it north to pawn three firearms he'd swiped from the family gun cabinet.

So Kathie sent her son above the Arctic Circle to spend some quality time with his father, who was selling food out of a converted school bus to truckers passing along the Trans-Alaska Pipeline. That operation would become a restaurant and service station and then the town of Coldfoot, Alaska, population: a dozen.

Working at his father's service station, Mackey learned to barter truck repairs for drugs. "Truck drivers are as bad of junkies as anybody you ever met," he says. "So I had access to just about every drug I could get my hands on." Mackey returned to Wasilla just before his eighteenth birthday and picked up his life of petty crime where he had left off—until one Saturday, when Kathie refused to bail her son out of jail.

When Lance got out, he headed to the Bering Sea, where he spent the next decade as a commercial fisherman on long-liners. Even then, Mackey would tell crewmates on fishing vessels—many of whom were from Mexico and had never heard of the Iditarod—that one day he would win the race that his father cofounded. "You ain't nothing as a musher unless you win the Iditarod," Mackey would recount his father saying.

By 1997, Mackey was living with Tonya in Nenana, Alaska, and both were addicted to cocaine. They occasionally used Amanda, Tonya's daughter from a previous marriage, as a designated driver. "She had a cushion so she could see over the wheel," Mackey says. "She thought it was cool as hell, being nine years old driving down the highway."

On June 2, 1998, Mackey's twenty-eighth birthday—and not long after he'd nearly gotten himself killed in a gun-filled bar brawl—he and Tonya decided to go cold turkey. On one night's packing, they moved 465 miles south to Alaska's Kenai Peninsula and left their drug habits behind. There, Lance and Tonya lived with Amanda and Brittney—Tonya's other daughter, then eight—beneath a tarp on the beach. A pup tent served as the master bedroom. For dinner, Tonya made a campfire and cooked flounder that the girls plucked from the sand. Lance started working for a construction crew and at a local sawmill. It was enough to make a down payment on a plot of land where he and Tonya built a timber house and stuffed the walls with clothing from the Salvation Army for insulation. With cocaine behind him, Mackey threw himself into a new addiction: breeding and raising sled dogs.

He had no money to buy the trim and powerful huskies that had already distinguished themselves in races, so he took in mutts from the street or adopted the castoffs of other mushers. Mackey accepted that his motley band of dogs would never be the sprinters of the canine world, so he decided to breed for other qualities, and that's when he met Rosie.

Rosie was a tiny female dog that once belonged to sprint racer Patty Moran. Moran decided that Rosie was too slow, so she sold her for a pittance to Rob Sparks, a musher who raced longer distances. When Sparks saw that Rosie refused to switch from a trot to a lope, he, too, decided that little Rosie was just too slow to race. At Sparks's offering, Mackey took Rosie out for a test drive. Sure, she wasn't fast, but Mackey saw something else: hook Rosie up to a sled harness and she'll trot, as Mackey puts it, until she bores a hole through the earth. He was glad to take her off Sparks's hands. His "trotting tornado," he called her.

Mackey bred Rosie with Doc Holliday, another husky that would never win a sprint but that yearned for nothing more than to run, eat, and run some more. From the union of Rosie and Doc Holliday, Mackey got Zorro.

Even elite-bred and trained sled dogs will regularly coast on a long run. That is, they'll slyly back off the pace when other members of the team are working hard. An experienced musher can tell when a dog is backing off because the rope—known as the "tug line"—that connects the dog to the sled's main line won't be perfectly taut. But Zorro was *always* pulling. From his very first race Zorro had to be restrained at the starting line and kept right on pulling even after the finish. Though Zorro was on the heavy side for a racing dog, "I told my brother Rick," Mackey says, "I'm breeding Zorro to every dog I own."

In 2001, Mackey picked a team from his band of rejects and hand-me-downs and put them together with Zorro—the lone dog he'd bred and raised—and entered the Iditarod. It took Mackey 12 days, 18 hours, 35 minutes, and 13 seconds to finish the race, good enough for thirty-sixth place. Zorro, not yet two years old, was the youngest dog in the entire field to complete the 1,100 miles, and he did it in great shape, barking and yanking the sled across the finish.

Mackey himself was less chipper. He had pushed through the pain of what multiple doctors had told him, erroneously, was an abscessed tooth. During the race he suffered from blurry vision, headaches, and blackouts. After the finish, he collapsed. Tonya took him straight to the hospital. The following week Mackey was in emergency surgery for throat cancer. It was the kind of surgery before which the doctor tells the patient to make sure there's nothing he'd regret having left unsaid to his wife and family. Mackey's normally staid father, Dick, was inconsolable.

Surgeons removed a grapefruit-sized tumor from Mackey's throat, along with the skin, muscle tissue, and salivary glands with which it was entangled. From then on, Mackey had to sip constantly from a water bottle or on the juice from a fruit cup in order to keep his throat moist enough that he could breathe. Radiation treatments that damaged Mackey's nerves left him with pulsing pain in his left index finger, so he went from doctor to doctor until he persuaded one to just cut the thing off.

Through it all, even when it seemed as though Mackey might not survive, Tonya kept his breeding plan going. At Mackey's direction, she bred Zorro with females in the yard. By the winter after his surgery, Mackey was well enough to return to work with sixty-six of Zorro's tongue-and-tail-wagging puppies to greet him.

Mackey returned to the Iditarod in 2002—with a feeding tube in his stomach—but withdrew after 440 miles. He skipped the next Iditarod, and for the next few years concentrated on raising and training Zorro's children and grandchildren. Mackey's training plan was tailored to his initial breeding strategy of mating the hardest-working dogs—the strategy foisted upon him because he couldn't afford the fastest dogs. Knowing he would never outrace his Iditarod competitors between checkpoints, Mackey developed what he calls his "marathon style," a technique that would transform long-distance dogsled racing. Rather than sprint between rest stops—as many successful mushers did at speeds up to 12, or occasionally 15 miles per hour—Mackey had dogs that were slower but would trot till they bored a hole in the earth. "At seven miles per hour, that's poking," Mackey says, "but if the dogs will go seven miles per hour for nineteen straight hours, then you're going to go places."

In 2007, Mackey started the Iditarod with a sixteen-dog team that consisted almost entirely of Zorro's progeny. Those that weren't directly from Zorro included his half-brother Larry, his nephew Battel, and Zorro himself. Just more than nine days later, with tears frozen to his face, Mackey passed beneath the burled arch in first place. "Life just changed," Mackey told his dogs. And so did dogsled racing.

Suddenly, Mackey's competitors wanted to copy the marathon style. Overnight his kennel went from home of cheap castoffs to a yard full of coveted bloodlines, with dogs worth four figures each, at least. (According to Tonya Mackey, Zorro's son Hobo was purchased by another musher and then "tossed all around Norway breeding, at a couple grand for each breeding.") In 2008, Mackey won the Iditarod again, his second of four straight. In a race a few weeks later, a drunk snowmobile driver rammed his team. Zorro had to be airlifted to Seattle with three broken ribs, a bruised lung, internal bleeding, and spinal damage so severe he couldn't stand.

Zorro survived, but a veterinarian ordered that he be retired from breeding and racing. Mackey built Zorro a doghouse in the yard, but quickly saw that his relentless runner would whine and tug at his chain if he was passed over while other dogs were taken out on a run. So Mackey built him the "condo on the hill," his fenced-in area in front of the house. "He's still the badass in the kennel," Mackey says. "He's my main man even though he doesn't run. He holds a very, very special place in my life and heart." And, more important, in the gene pool in his front yard.

The idea that dogs can be bred to win a race is no revelation. Darwin himself marveled at the ability of dog breeders to cultivate almost any trait they wanted. Breeding for speed in racing whippets has been so intense that over 40 percent of the dogs in the top division have what is normally an exceedingly rare myostatin gene mutation (the "Superbaby" mutation).

In the late nineteenth and early twentieth centuries—particularly during the Klondike Gold Rush—when the seaports and rivers of Alaska were frozen solid, sled dogs were the main source of transport for everything from mail to gold ore. Breeding for strength, endurance, and resistance to cold proceeded in earnest, until snowmobiles came into fashion. When dogsled racing gained popularity with the rise of prize money following the first Iditarod in 1973, breeding for athleticism became a serious business. Pointers, salukis, and a bundle of other breeds were mixed into a genetic stew that had traditionally included Alaskan malamutes and Siberian huskies. It worked.

The winners of the first two Iditarod races took more than twenty days to finish. Two decades of breeding later, mushers were finishing in half that time. Alaskan huskies morphed into athletes unique on the planet. Even before training, an elite Alaskan husky can move four to five times as much oxygen as a healthy, untrained adult man. With training, top sled dogs reach a VO_2max about eight times that of an average man, and more than four times higher than a trained Paula Radcliffe, the women's marathon world record holder.

Sled dogs were bred for everything from a voracious appetite—they eat ten thousand calories a day during the Iditarod—to webbed toes ideal for traveling atop snow, to a pulse rate that settles quickly at a moment's rest. Perhaps the most remarkable bit of biology bred into Alaskan huskies is the ability to adapt almost instantly to exercise. As in humans, when sled dogs start training, they deplete the energy reserves in their muscles, undergo an increase in stress hormones, and damage cells. Human athletes experience this as fatigue and soreness, and must rest to allow the body to adapt to the exercise before coming back to training or racing. But the best sled dogs adapt *on the run*. Whereas humans have to alternate exercise and rest to get fit, premier Alaskan huskies get fit while barely stopping to recuperate. They are the ultimate training responders.

In 2010, Heather Huson, a geneticist then studying at the University of Alaska, Fairbanks—and a dogsled racer since age seven—tested dogs from eight different racing kennels. To Huson's surprise, Alaskan sled dogs have been so thoroughly bred for specific traits that analysis of microsatellites—repeats of small sequences of DNA—proved Alaskan huskies to be an entirely genetically distinct breed, as unique as poodles or labs, rather than just a variation of Alaskan malamutes or Siberian huskies.

Huson and colleagues discovered genetic traces of twenty-one dog breeds, in addition to the unique Alaskan husky signature. The research team also established that the dogs had widely disparate work ethics (measured via the tension in their tug lines) and that sled dogs with better work ethics had more DNA from Anatolian shepherds—a muscular, often blond breed of dog originally prized as a guardian of sheep because it would eagerly do battle with wolves. That Anatolian shepherd genes uniquely contribute to the work ethic of sled dogs was a new finding, but the best mushers already knew that work ethic is specifically bred into dogs.

"Yeah, thirty-eight years ago in the Iditarod there were dogs that weren't enthused about doing it, and that were forced to do it," Mackey says. "I want to be out there and have the privilege of going along for the ride because they want to go, because they love what they do, not because I want to go across the state of Alaska for my satisfaction, but because they love doing it. And that's what's happened over forty years of breeding. We've made and designed dogs suited for desire."

Several mushers I spoke with suggested that sled dogs may have maxed out their physiological capacity and are no longer getting faster or hardier, and that the improvement in race times is now entirely down to how long the dogs are eager to pull without rest. "The dogs are in control," says Eric Morris, the biochemist and musher. "That's why we breed dogs that want to do it . . . it's something I had to learn through trial and error, and time, and speaking and working with other mushers, to find out what all the great ones know. The great mushers know how to breed a dog that has drive and the desire to pull, and then they foster and develop that desire."*

Scientists who breed rodents for their desire to run have proven that work ethic is genetically influenced. One of the leaders in that field has been Theodore Garland, a physiologist at UC Riverside. For more than a decade, he has been offering mice a wheel that they may hop on or shun at their discretion.

Normal mice run three to four miles each night. Garland took a group of average mice and separated them into two subgroups: those that chose to run less than average each night, and those that chose to run more than average. Garland then bred "high runners" with other high runners, and "low runners" with other low runners. After just one generation of breeding, the progeny of the high runners were, of their own accord, running even farther on average than their parents. By the sixteenth generation of breeding, the high runners were voluntarily cranking out seven miles each night. "The normal mice are out for a leisurely stroll," Garland says. "They putz around on the wheel, while the high runners are really running."

When mice are bred for endurance capacity—not voluntary running, but when they are forced to run as long as they physically can—successive generations have more symmetrical bones, lower body fat, and larger hearts. In his voluntary-runner breeding program, Garland

*I experienced Alaskan husky desire the hard way. On my first and only dogsled trip, in 2010 over the frozen Boundary Waters of Minnesota, my lead husky was a retired racer who was—I later learned—one of Zorro's sons. I needed about one hundred meters of forceful braking to get the dogs to pause atop a frozen lake. But as soon as I eased up on the brake and looked to the side, the team bolted. I was tossed off the sled and had to chase it for a quarter mile, until the sled got wedged between trees on a tiny frozen island. Lucky for me, as I'm pretty sure I would've given up well before Zorro's progeny.

saw body changes, "but at the same time," he says, "clearly the brains are very different." Like their hearts, the brains of the high runners were larger than those of average mice. "Presumably," Garland says, "the centers of the brain that deal with motivation and reward have gotten larger."

He then dosed the mice with Ritalin, a stimulant that alters levels of dopamine. Dopamine is a neurotransmitter, a chemical that conveys messages between brain cells. The normal mice, once doped, apparently derived a greater sensation of pleasure from running, so they started doing it more. But the high runners, when doped, did not run more. Whatever Ritalin does in the brains of normal mice is already occurring in the brains of the high-running mice. They are, quite literally, running junkies.*

"Who says motivation isn't genetic?" Garland asks, rhetorically. "In these mice, it's absolutely the case that motivation has evolved."

Researchers around the world have begun to explore locations on the genome that differ between marathon mice and their normal counterparts, and specifically to home in on genes related to dopamine processing that might impact the sense of pleasure or reward a mouse gets from a particular behavior.

Of course, they aren't doing this simply to understand why rodents want to run. The ultimate goal is to learn about human gym rats.

Pam Reed was up on top of the parking garage at LaGuardia Airport in Queens, again. Her flight out of New York City was delayed, and she was never one for sitting still. While disgruntled travelers jostled for electrical outlets and cushioned seats, their bags trundling behind

---

*Anything that can be bred for must have a genetic component, or else the breeding would not work. Researchers have managed to successfully breed rodents for some bizarre traits, like voluntary gnawing on their own toes. Just as with voluntary running, if toe-masticating mice are bred with one another, generations down the line they produce progeny that will completely bite their toes off.

them, the fifty-one-year-old Reed popped in her earbuds and headed for the top deck of the parking garage.

She breathed in the thick summer air. Reed stashed her luggage in a corner and started running. Immediately, a placid calm dripped through her body. For a good hour, she ran around and around in tight circles, each lap no more than 200 meters. It certainly wasn't because she needed the fitness.

Just the previous day, Reed had finished the U.S. championship Ironman triathlon in New York City in 11 hours, 20 minutes, and 49 seconds, good enough to qualify for the world championship in Hawaii. A week before that, she participated in a relay race in which her leg consisted of eight continuous hours of circling a track. Two weeks before that, she spent 31 hours running en route to becoming the second female finisher at the 2012 Badwater Ultramarathon, a 135-mile race that starts in Death Valley, and that Reed has won, twice.

Reed's flight out of LaGuardia eventually left, and the next weekend she completed the Mont-Tremblant Ironman in Québec in 12 hours, 16 minutes, and 42 seconds. The weekend after that, she had "only a marathon," she says, never mind that it was through the Tetons, in her home of Jackson Hole, Wyoming.

This isn't some masochistic running binge, it's life for a woman who once ran three hundred miles without sleeping, and in 2009 spent six days running 491 laps around a drab one-mile loop in a park in Queens.

When she was an eleven-year-old in Michigan, Reed was smitten by her first sports love while watching the 1972 Olympics on TV: gymnastics. "I was obsessed," Reed later wrote in her autobiography, *The Extra Mile*. "I practiced gymnastics every minute that I could, in the basement, off the couch, wherever I happened to be." In high school, Reed turned to tennis, and, as usual, threw herself into it the way a Navy SEAL throws himself out of a plane—with gusto. Part of her training was a minimum of one thousand sit-ups a day. She went on to play varsity tennis at Michigan Tech. When she later moved to Arizona—she owns and directs the Tucson Marathon—she worked as an aerobics instructor so that she could have access to the health club's pool. Naturally (for Reed), she fell in love with her second husband as the pair trained together for an Ironman triathlon. Reed has often wondered about the source of her relentless drive to be in motion.

Her father was tireless. He used to rise at 3:30 A.M. to head to work at an iron mine, and when he returned home in the afternoon he would go straight to building an addition to the house or tinkering on the car. According to family lore ("absolutely true," Reed says), her grandfather Leonard once got into an argument at a family gathering in Merrill, Wisconsin, and stormed out in a huff. He kept walking. The entire three hundred miles back home to Chicago.

"Running for three hours every day might put some people in the hospital," Reed writes in her book, while noting that she finds peace of mind in extreme activity. "I am certain that *not* running for three hours every day would very quickly make me ill. . . . While nobody's forcing me to do this, it's not really a choice, either. There's something in my nature that makes it really hard for me to sit still . . . being temperamentally attuned to perpetual motion makes me pretty uncomfortable on long car trips or in sedate social settings." (Reed's son Tim contrasts himself to his mother: "I only like to run for maybe two or three hours max.") One of Reed's current goals is to set the women's world record for running across America, which she plans to do at a pace of two marathons a day.

"When I don't do this," Reed says—and by "this" she means running three to five times a day—"I feel horrible. I had C-sections, and three days after them I was running. . . . It's who I am. I totally love it. As I get older, I have to say, I can sit still a bit longer, but it's not comfortable."

In her book, Reed astutely ponders whether she might be the human version of the rodents from an experiment at the University of Wisconsin in which mice bred for voluntary running were restricted from running, and then had their brain activity measured. Brain circuitry similar to that which is active when humans crave food or sex, or when addicts crave drugs, was activated in the high-running mice that were denied the chance to run, and they became agitated. The researchers presumed that when the mice were deprived of running their brain activity would decline. Instead, it went into overdrive, as if the mice needed exercise to feel normal. The longer the distance a particular mouse was used to running, the more frenetic its brain activity became when it was made to sit still. As with Garland's mice, these rodents were genetic junkies for exercise.

Pam Reed is an outlier by any measure. But a seemingly compulsive drive to exercise is hardly unique among distinguished athletes. Consider Ethiopian Haile Gebrselassie, who has set twenty-seven distance running world records: "A day I don't run, I don't feel good," he says. Or Floyd Mayweather Jr., the undefeated boxing champion, who has been known to jolt awake in the middle of the night and force his bloated entourage to meet him at the gym for a workout. Or Steve Mesler, a member of the 2010 Olympic four-man bobsled team that won the first U.S. gold in sixty-two years. He retired afterward, but says he "feels anxious" when he takes a break from working out even now. Or Ironman triathlete Chrissie Wellington or high jumper Stefan Holm, both of whom claim addictive personalities that they channeled to their training.

Or Herschel Walker, best known as the 1982 Heisman Trophy–winning running back and twelve-year NFL veteran. Now fifty-one, Walker is 2-0 as a professional mixed martial artist. Walker has trained in ballet, taekwondo (he's a fifth-degree black belt), and, in 1992, was an Olympic bobsled pusher. Most indicative of Walker's drive to be active, though, is the workout regimen he started at age twelve, before he was involved in organized sports, and which he has continued every day since. "I would start doing sit-ups and push-ups at seven P.M.," he says, "and go until eleven. It was every night, on the floor. It was about five thousand sit-ups and push-ups." These days, Walker says he "only" does 1,500 push-ups and 3,500 sit-ups a day—in sets of 50 to 75

push-ups and 300 to 500 sit-ups or crunches—but he also has his martial arts training.

Walker says the push-ups and sit-ups routine will remain, even after
he stops competing. "It has nothing to do with my competitions," he
says. "It becomes a drug, or a medicine. Even if I'm sick, I do it. It's like
there's something saying, 'Herschel, you gotta get up. You gotta do it.'"

Variations in the brain's dopamine system make certain individuals
more likely to feel reward when using particular drugs, and they are
more likely to become addicted. Is it possible that, like sled dogs and
lab mice, some people are biologically predisposed to get an outsized
sense of reward or pleasure from being constantly in motion?* All
sixteen human studies conducted as of this writing have found a large
contribution of heredity to the amount of voluntary physical activity
that people undertake.

A 2006 Swedish study of 13,000 pairs of fraternal and identical
twins—fraternal twins share half their genes on average, while identical twins essentially share them all—reported that the physical activity
levels of identical twins were twice as likely to be similar as those of
fraternal twins. That study used a survey to measure physical activity,
though, and people chronically overestimate their own physical activity levels. But another, smaller study of twin pairs that used accelerometers to measure physical activity directly found the same difference
between fraternal and identical twin pairs. The largest study, of 37,051
twin pairs from six European countries and Australia, concluded that
about half to three quarters of the variation in the amount of exercise

*In her engrossing book *Gifted Children: Myths and Realities*, psychologist Ellen Winner coined
the phrase "rage to master" to describe one of the primary qualities of gifted children. She
describes it as intrinsic motivation and "intense and obsessive interest." In a sentence that
seems as if it were made to describe Tiger Woods or Mozart, she writes: "The lucky combination
of obsessive interest in a domain along with an ability to learn easily in that domain leads to
high achievement."

people undertook was attributable to their genetic inheritance, while unique environmental factors, like access to a health club, had a comparatively puny influence.

It is entirely clear that the dopamine system *responds* to physical activity. This is one reason that exercise can be used as part of treatment for depression and as a method to slow the progression of Parkinson's disease, an illness that involves the destruction of brain cells that make dopamine. And there is evidence that the reverse is true as well, that physical activity levels respond to the dopamine system. Several lines of scientific evidence have begun to implicate genes that control dopamine.

Particular versions of dopamine receptor genes have been associated with higher physical activity and lower body mass index. Multiple studies—including a meta-analysis of all published studies—have also replicated the finding that one of those variants, the 7R version of the DRD4 gene, increases an individual's risk for attention deficit hyperactivity disorder, or ADHD. Tim Lightfoot, director of the Sydney and J. L. Huffines Institute for Sports Medicine and Human Performance at Texas A&M, has authored papers on voluntary physical activity in rodents and humans, and he sees a connection between ADHD, exercise, and dopamine genes. "The high active mice we bred in the lab," Lightfoot says, "they mimic ADHD kids, at least as far as the dopamine system goes. . . . They're low on [a particular kind of] dopamine receptors, and if you can drive the amount of dopamine up, their physical activity decreases."

Ritalin drives dopamine up in hyperactive children, and their activity decreases. Obviously, this is a good thing for a child who is having difficulty sitting still in school. But, Lightfoot suggests, it might have unintended consequences. "These may be kids that have a very strong drive to be active, and maybe we're blunting it with medications."

"Our society is so scared right now of kids being fat," Lightfoot continues. "Well, what if we're putting some of these kids on drugs that actually may be contributing to this by driving their activity levels down?" In any case, that's exactly what happened in Lightfoot's mice.

A set of scientists have proposed the controversial idea that hyperactivity and impulsivity may have had advantages in the ancestral state of man in nature, leading to the preservation of genes that increase ADHD risk. Interestingly, the 7R variant of the DRD4 gene is more common in populations that have migrated long distances, as well as those that are nomadic, compared with settled populations.

In 2008, a team of anthropologists genetically tested Ariaal tribesmen in northern Kenya, some of whom are nomadic and some recently settled. In the nomadic group—and only in the nomadic group—those with the 7R version of the DRD4 gene were less likely to be undernourished. One of several hypotheses the researchers offered: "It might also be that higher activity levels in [the 7R] nomads are translated into increased food production." In other words, it could be that carriers of that version of the gene are harder workers when it comes to physical activities.

"One of the issues with our field is when we've looked at activity, and what controls activity, we've forgotten that we know very clearly there are biological mechanisms that actually influence people to be active or not," Lightfoot says. "You *can* have a predisposition to be a couch potato."

Quite obviously, as is the case with Kenyan children, the necessity of transportation by foot and the aspiration for a better life can have profound influences on physical activity levels. But those environmental factors do not exclude the significant contribution of genetics that has shown up in every study ever conducted on the heritability of voluntary physical activity.

Those consistent findings are reminiscent of a famous quote by Wayne Gretzky, the greatest hockey player in history: "Maybe it wasn't talent the Lord gave me, maybe it was the passion."

Or maybe the two are inextricable.

Even as it has been demonstrated in study after study that genetic inheritance influences physical activity, scientists are only beginning to discern the specific biological processes that play a role. Plus, every scientist knows full well that extreme environments can dramatically alter how much an individual trains. While dopamine plays a role in the drive to be in motion, there are certain, more obvious enticements.

When Floyd Mayweather Jr., renowned for his furious training, dropped by the *Sports Illustrated* office in 2007, fresh off his victory over Oscar De La Hoya, he described an unhappy period in his past when he was constantly concerned about money. "But I'm happy now," he said with a mile-wide smile, referring to the $25 million he made for the fight.

All told, the tangle of nature and nurture is so complex as to prompt the question: can there possibly be any practical use at all for genetic testing in sports right now, in the present day?

The answer, despite all the complexity: absolutely.

15

The Heartbreak Gene
Death, Injury, and Pain on the Field

I wasn't there that day, February 12, 2000, in the desiccated winter air at the indoor track at Evanston Township High School. I had graduated and was off running at college. But my brother was a freshman on the team and my father was there too, videotaping. He was among the spectators in the aluminum bleachers standing up for a better look when my friend and former training partner, Kevin Richards, went down.

It was not unusual for a bone-tired runner to crumple to the ground after a hard race. But it had never happened to Kevin before. His teammates knew him to address his aches in silence, and always standing up. He embraced the pain of a race, and scorned the practice of lying down in exhaustion. "I love being sore," he once said. "It feels like you did something."

Normally, a fallen runner draws only slight curiosity from a track-savvy crowd. But Kevin was a state champion, and the dusty, green rubber floor of the track was no place for a state champion to be lying on his back, shuddering.

Kevin's mother, Gwendolyn, had sensed something was wrong with him that morning when he overslept. He never overslept on race day. She thought he must be getting sick, so she asked him not to go. But Dan Glaz of Amos Alonzo Stagg High School was in town to race the mile. Glaz was one of the best runners in Illinois. He would become a state champion and earn a scholarship to Ohio State.

Kevin was a junior, and he was getting recruiting packets too. In addition to being one of the top half-milers in Illinois, Kevin was an honors student, and would be the first in his family of Jamaican immigrants to attend college. He had told me—often while I was struggling to breathe on our runs—that he wanted to be a video game designer, and Indiana University was atop his college wish list. On this day, he wasn't about to miss a chance to run against Glaz, a potential future Big Ten rival.

Gwendolyn, who worked at a nursing home, had been reluctant to bank on Kevin's speed, so she attended financial aid seminars to figure out how to pay for college, until Kevin told her to stop. "You aren't paying a penny for me," he said, and then turned his back and walked away.

Moments before he crumpled to the floor, Kevin had been flying through the final surge, in pursuit of Glaz. There were other runners in the race, but it had become a duel between Kevin and Glaz. The duo had already lapped the field. With two laps to go, Glaz opened a gap, but as the bell clanged hollowly, signaling the final lap, Kevin reached down. He came steaming back around the final curve, swallowing ground and slicing into Glaz's lead with each gaping stride. He ran out of room, just barely, and finished on Glaz's shoulder, in second place.

Kevin walked a few exhausted steps past the finish line. As coach David Phillips came to lend a supporting arm, Kevin slipped through his grasp to the floor and started to shake.

Bruce Romain, the head athletic trainer, had seen nearly one hundred seizures in his career. He knelt beside Kevin and took his pulse. It was racing. He squeezed Kevin's hand. Kevin did not squeeze back, but continued, like a fish washed ashore, to quake and heave and force air out of his mouth. Bits of saliva frothed over his lower lip with each labored breath.

A fireman among the spectators called the paramedics. Within minutes of Kevin's collapse, emergency medical technicians raced into the field house to help Romain give him mouth-to-mouth resuscitation. Kevin gave a mighty suck, and then exhaled a long, lackadaisical sigh. He stopped breathing.

Romain looked across Kevin's body at a medic. The eyes of the two professional rescuers locked. "Oh, shit," Romain blurted, as Kevin's pulse disappeared. A medic rushed back to the rig for defibrillator paddles. Romain and the other medic continued furiously to give Kevin CPR. One of them acted as Kevin's lungs, blowing oxygen-rich air into his mouth. The other was his heart, pushing down on his chest to force the oxygenated blood to flow through his body. But CPR could only buy some time. It could not make Kevin's heart beat again. Like a car in need of a jump start, only a machine could save him now.

Somewhere on that last lap, the electrical signals that cued Kevin's heart to pump had begun to misfire horribly. Rather than contracting and relaxing rhythmically, Kevin's heart trembled, like Jell-O on a shaken tray. His left ventricle, the chamber that takes oxygenated blood from the lungs and squeezes forcefully, sending it hurtling through the body, had malfunctioned, causing a circulatory traffic jam. Blood backed up in the capillaries in his lungs—vessels so narrow that red blood cells have to move through them in single file—while the water in Kevin's bloodstream pushed through the capillary walls and seeped into the tiny air sacs in his lungs. Water occupied the space where oxygen should have been. Kevin began drowning in his own body's water.

The medic returned with the defibrillator paddles. They would try to jolt Kevin's heart back into a normal rhythm by jarring it with electricity. They meant to shock him back to life, and they hoped they could do it sooner rather than later. Of all the times Kevin had been measured by the clock, these next few minutes on the track would be the most critical of his life. In the time it took him to run a mile, Kevin's brain cells would begin to die in droves, in the poisonous, oxygenless environment of his own head.

One of Kevin's teammates paced back and forth near the finish line murmuring, "No way. He's too strong." Romain backed away, stunned. He told one of the assistant coaches to call Gwendolyn at her work. By the time she arrived, her son was being loaded into an ambulance. She forced her way into the passenger seat. A medic pulled down a shade so she could not see her son being worked on in the back.

When they arrived at Evanston Hospital, Gwendolyn sat in the waiting room, passing the longest minutes of her life. Then a chaplain came to meet her. "I know he's dead! Just tell me he's dead!" she screamed. Then she fainted.

Kevin was dead. He had died at the track.

Somewhere amid the three billion base pairs—the chemical compounds that form the rungs of the twisting DNA ladder—Kevin had a single misspelling in his genetic code. That's like a single typo in a string of letters vast enough to fill thirteen complete sets of the *Encyclopaedia Britannica*.

Kevin's genetic mutation could have been in any one of billions of locations. One particular spot would have caused him to have muscular dystrophy, while another would have left him colorblind. Many, many other locations would have had no discernible impact at all, as is the case with most of the mutations that each one of us carries around every day. But Kevin's mutation occurred at the precise rung of the DNA ladder to draft the biological blueprint for a broken heart.

Kevin had hypertrophic cardiomyopathy, or HCM, a genetic disease that causes the walls of the left ventricle to thicken, such that it does not relax completely between beats and can impede blood flow into the heart itself. About one in every five hundred Americans has HCM, though many will never exhibit serious symptoms. According to Barry Maron, director of the Hypertrophic Cardiomyopathy Center at the Minneapolis Heart Institute Foundation, HCM is the most common cause of natural sudden death in young people. And it's easily the most common cause of sudden death in young athletes.

According to statistics that Maron has compiled, at least one high school, college, or pro athlete with HCM will drop dead somewhere in the United States every other week. Some of them will be famous, like Atlanta Hawks center Jason Collier, or San Francisco 49ers offensive lineman Thomas Herrion, or Cameroonian soccer pro Marc-Vivien Foé. Most, though, will be like Kevin Richards—teenagers, just on the verge of becoming.

In those people, the muscle cells of the left ventricle are not stacked neatly like bricks in a wall, as they should be, but are all askew, as if the bricks had instead been dumped in a pile. When the electrical signal that cues the heart to flex travels across the cells, it is liable to bounce around erratically. Intense athletic activity can trigger this short-circuit, which is especially dangerous during competition, when an athlete straining his body will not respond to the early signs of danger.

For the nation's most pressing health problems—diabetes, hypertension, coronary artery disease—exercise is a miraculous medicine. But people with HCM can be at increased risk of dropping dead precisely *because* they exercise.

Eileen Kogut, for example, had long known that something dangerous ran in her family. When Kogut was twenty-one, in 1978, her fifteen-year-old brother, Joe, was playfully roughhousing with their brother Mark at the dinner table when Joe dropped dead. The autopsy report listed the cause of death as "idiopathic hypertrophic subaortic stenosis"—essentially, a heart that is enlarged for unknown reasons. "Joe was the youngest of seven siblings," Eileen says. "His death was incredibly devastating to our family." So Mark, who carried the memory of his little brother dying right in front of him, started working out every day, just in case he had a faulty heart, like Joe. Mark was on a treadmill at the YMCA in Lansdowne, Pennsylvania, in 1998, when he collapsed and died. The cause of death, again: an enlarged heart of unknown cause. Mark was thirty-seven. He left behind a wife and three young sons.

HCM is passed down in what's called "autosomal dominant" fashion, which simply means that there is a 50-50 chance—a coin flip—that a parent with the culprit gene will pass it to a child.

Eileen Kogut eventually learned that HCM was what took her brothers, and in 2008 she decided to look into her own DNA.

Just across the Charles River from Boston's Fenway Park is another structure of brick and steel. But rather than flags commemorating World Series wins, snaking down three stories of the exterior of this building are two winding metal ribbons, an artistic depiction of the DNA double helix.

Inside the building—the Harvard-affiliated Partners HealthCare Center for Personalized Genetic Medicine—geneticist Heidi Rehm directs the Laboratory for Molecular Medicine. Rehm and her lab staff identify new HCM mutations by the week. In the early 1990s, it was thought that HCM came from any one of seven different mutations on a single gene, the MYH7 gene, which codes for a protein found in heart muscle. By the time I visited Rehm's lab in 2012, there was a database that included 18 different genes and 1,452 different mutations (and counting), any one of which can cause HCM. Most of the mutations are in genes that code for proteins found in heart muscle, and around 70 percent of people with HCM have a mutation on just one of two specific genes. (To make matters extremely complicated, though, two thirds of the different HCM mutations are "private mutations." That is, each one has only been identified in a single family.) The most common cause of HCM is a DNA spelling error known as a "missense" mutation. A missense mutation occurs when a single letter is swapped in the DNA code, but in such an important place that it changes the amino acid that goes into making the resulting protein.

HCM mutations can occur randomly in someone with no family history of the disease, but most HCM gene variants are passed from parents to children. Some, however, don't make it down family lines. One particularly dangerous HCM gene variant only ever appears as a spontaneous mutation in a single individual in a family. "That's because it's reproductive lethal," Rehm says. "No one ever survives to an age to reproduce and pass it on."

Other mutations can be so mild as to go entirely unnoticed over a lifetime, like the "Trp-792 frameshift," which sounds like it's out of an NFL playbook, but is actually a mutation found specifically in Mennonite people.

In most instances, though, it's difficult to tell whether a particular mutation puts an HCM patient at risk of sudden death. In Kevin's case, the disease was only diagnosed after he died and his heart was examined. Kevin's autopsy showed that his heart was a gargantuan 554 grams. An average adult male heart is around 300 grams. Kevin had no obvious signs of disease, other than that he had once been told he had a heart murmur. But so had I, and so have hordes of athletes who have been at the flat end of a stethoscope. As with any muscle, the heart gets stronger with exercise, and athletes often have nondangerous heart murmurs that go away when they're out of shape.*

Given her family history, Eileen Kogut had all her children's hearts checked regularly from the time they were little. Her son Jimmy, who played basketball and lifted weights, had occasionally complained of shortness of breath. He was told he had asthma, a common but dangerous misdiagnosis for someone with HCM, because asthma inhalers can prompt lethal heart rhythms in HCM patients. In 2007, as he was getting set to start junior year at the University of Pittsburgh, Jimmy had a genetic test and learned that he has one of the most common HCM mutations, on a gene that helps to regulate heart contraction. Like his hazel eyes and freckles, he got it from Eileen. With the family mutation identified, Eileen decided to have her other children,

---

*A troubling trend in high school sports is the increasing number of states that are allowing health care providers who have little or no cardiovascular training—and thus no chance of identifying a dangerous heart murmur—to conduct the preparticipation screening of athletes. In 1997, eleven states allowed chiropractors, herbalists, or other nonphysicians to perform the exams. By 2005, that number had increased to eighteen states, with three states—California, Hawaii, and Vermont—allowing high schools to decide who can perform the exams.

Kyle, then eighteen, Connor, then sixteen, and Kathleen, then twelve, tested, even though they weren't showing symptoms. In March 2008, she took the kids for genetic screening and prayed that she hadn't passed the mutation to any more of her children.

But the tidings were bad. Connor and Kathleen both came up positive. "I was devastated," Eileen says. "I don't know what I expected. I expected to hear good news. It was not an easy pill to swallow . . . I was angry at the lab. I was not coping well. I just thought, 'Why did I ever do this? They're young, what was I thinking? It's going to ruin their childhood.'"

Cardiologists who study HCM recommend that people with the disease abstain from extremely rigorous activity, because the increase in adrenaline may spark a deadly heart rhythm. After the diagnosis, Jimmy underwent surgery to have a defibrillator implanted in his chest. About the size of a matchbox, the tiny device has wires that reach into the heart and stand guard, waiting for an abnormal heart rhythm. If one is detected, the defibrillator automatically fires an electrical shock to jolt the heart back to a normal pattern. Jimmy returned to college life as usual, minus the basketball. And weight lifting was restricted to nothing overhead or that could stress his left side so much that it might damage the defibrillator wires.

Ultimately, Eileen overcame her dismay and is glad she had her children tested, even though it meant certain lifestyle changes. As she had learned in the cruelest way, the only outcome worse than losing one brother is losing two. And the only fate worse than that would be losing two brothers and a child. Says Rehm, "I got totally hooked on this area of genetics because it really is an area where you can make a difference in patients' lives—to be able to figure out the cause of their HCM, and predict it in other family members. Sometimes you get bad outcomes, sometimes you get good outcomes, but at least you can understand it and predict it."

A definitive determination of HCM is particularly important in athletes, because the most conspicuous sign of HCM is an enlarged heart, which is normal for athletes. It often takes a true HCM expert—of which there are precious few in the world—to tell whether the enlargement is the result of the athlete's training or a sign of HCM. Martin Maron, Barry Maron's son, a cardiologist at Tufts Medical Center in Boston and an expert on sudden death in athletes, says that the specific enlargement depends on the sport the athlete plays. Cyclists and rowers, for example, have enlargements in the heart chambers and walls from their training, whereas weight lifters have thicker walls but not chambers. Each sport has its signature pattern.

In a normal heart, the wall that divides the heart chambers is usually thinner than 1.2 centimeters, and the left ventricle chamber is typically smaller than 5.5 centimeters across. If either the wall or chamber is greatly enlarged, it's a sign of disease. But if there is only some enlargement—a wall between 1.3 and 1.5 centimeters, and a chamber between 5.5 and 7 centimeters—then "that's a gray zone for athletes," Maron says. That is, the enlargement could be due either to training or to disease, and some athletes who are in the gray zone are cleared to play sports on the assumption that their large hearts are a training adaptation, only to then drop dead on the field. If, instead, the athlete is genetically tested and revealed to have a known mutation for HCM, no more gray zone.

This is one area where personalized genetic testing is making an impact on athletes in the present day—although they aren't always eager to take advantage of it.

In 2005, center Eddy Curry was leading the Chicago Bulls in scoring when he was sidelined with an irregular heartbeat. Curry missed the end of the season and the entire playoffs while he was being evaluated.

At the suggestion of Barry Maron, the Bulls—hoping to avoid a situation where Curry might die in front of television cameras, as the NCAA's reigning scoring and rebounding leader Hank Gathers did during a game in 1990—added a genetic testing clause to the $5 million contract offer that was on the table for Curry. If a test showed that

Curry had a known HCM gene variant, the Bulls would not allow Curry to play, but would pay him $400,000 per year for the next fifty years. Curry refused the test, and the Bulls subsequently traded him to the Knicks. "As far as DNA testing, we're just at the beginning of that universe," Curry's attorney, Alan Milstein, told the Associated Press. "Pretty soon, though, we'll know whether someone is predisposed to cancer, alcoholism, obesity, baldness and who knows what else . . . Hand that information to an employer and imagine the implications."

Today, the situation would be different. After thirteen years of haggling over genetic privacy, the U.S. Congress passed into law the Genetic Information Nondiscrimination Act of 2008, or GINA. The law took effect in late 2009, and barred employers from demanding genetic information, and both employers and health insurance companies from discriminating based on genetic information. (GINA does not, however, prohibit discrimination by providers of life, disability, and long-term care insurance.)

Plenty of athletes, even knowing they carry a dangerous mutation, choose to continue playing. In a 2009 moment immortalized on You-Tube, Anthony Van Loo, then a twenty-year-old defender on the Belgian soccer team SV Roeselare, crumpled to the pitch like a marionette whose strings had been cut. Van Loo was in cardiac arrest. Seconds later, he jerked violently and then sat up, as if nothing had happened. Van Loo's implanted defibrillator had fired and literally yanked him back from death's door. He was lucky, as implantable defibrillators aren't built to withstand the wear and tear of vigorous sports.

Whether to let an athlete with HCM participate in sports is a dilemma for doctors, who are frequently left guessing if their particular HCM patient is one of those who is at risk of sudden death, or one of those who will live to ninety with no serious symptoms.

Certain HCM mutations are known to be more dangerous than others, but it's an inexact science. "I see some kids, and they don't have a family history of death and they don't have symptoms or a very thick heart, and I don't think a lot of them are at great risk," says Paul D. Thompson, a cardiologist at Hartford Hospital and a competitor in the 1972 U.S. Olympic marathon trials. "I usually say to them, 'I don't think you're at great risk, but I have to sleep at night, and I can't take a chance with you, so I'm prohibiting you.' For some acne-stained seventeen-year-old who's accepted at that high school because he's a good linebacker, to tell him that's gone is a load."

It's better, though, than the linebacker himself being gone. When I went home for my friend Kevin's funeral, I visited the indoor track where he died. One of the white lines that demarcates a track lane was covered in penned messages: "LUV YA 4 LIFE"; "HOPE TO SEE YOU ON THE OTHER SIDE"; "WHEN THE TIME COME, YOU GOING TO LET US ALL KNOW WHY YOU DIED." When I visited again, a year later, the messages were there, in the floor with Kevin's sweat and dreams, but they were invisible beneath a fresh coat of paint.

Kevin never knew he had a time bomb inside his chest. But what if he had? At his funeral, friends emphasized that he died doing what he loved. Kevin *did* love to race. But he loved other things too, like computers. Racing might have been his scholarship ticket, but I have no doubt that he would have stopped running and eagerly rechanneled his competitive energy elsewhere. For me, there is scant solace in the poetic detail that he died running.

While the issue of whether to preemptively restrict an athlete is beset with emotional and legal barbs, cardiologists agree that when an athlete is clearly at risk of dropping dead on the field, the recommendation should be to avoid the field. (Though some athletes ignore the advice and play anyway.) But what if the athlete was just at risk of damage? Sports are inherently risky. Like flying fighter jets, no one participates for too long without an injury. But what if scientists could tell that some athletes are at greater risk than others?

Right now, they are beginning to be able to do just that, as researchers probe genes associated with some of the most high-profile medical risks in all of sports.

On a brisk November afternoon in Manhattan, Ron Duguay had just finished several hours of cognitive testing when he settled into a chair overlooking Park Avenue South to await news from Dr. Eric Braverman. Beginning in 1977, Duguay played twelve seasons in the NHL, primarily as a center with the New York Rangers. Duguay was a good player—he made the '82 All-Star Game—but was better known as hockey's rock star.

Duguay didn't wear a helmet, and the curly brunette locks that fluttered behind him when he skated made him a sex symbol in the 1980s. Even today, in his fifties and married to erstwhile supermodel Kim Alexis, Duguay's hair is thick and curly, and he is friendly and easy to talk to. In Braverman's office, though, he's nervous. He fiddles with his gleaming Rangers pinkie ring when he mentions that friends often tell him he should write a book about his hockey days. "I'd have to call up my teammates," he says. "There's a lot I can't remember." That's why Duguay is here. He thinks he suffered undiagnosed concussions during his career, and he knows he took scores of lesser hits to the head from sticks, elbows, and the occasional puck.

Braverman appears and flatly tells Duguay that he flunked three of the tests meant to gauge his memory and brain processing speed. "He's a mess compared to his old self," Braverman says.

As part of the testing, Braverman also ordered a genetic test to see what versions Duguay has of a gene known as apolipoprotein E, or ApoE. Duguay's grandmother died from Alzheimer's disease, and another family member has been having memory problems. Studies of Alzheimer's patients indicate that a particular version of the ApoE gene substantially increases an individual's risk of getting the disease.

The gene comes in three common variants: ApoE2, ApoE3, and ApoE4. Everyone has two copies of the ApoE gene, one from Mom and one from Dad, and a single ApoE4 copy increases the risk of Alzheimer's threefold. Two copies increases the risk eightfold. Around half of Alzheimer's patients have an ApoE4 gene—compared with a quarter of the general population—and those who do tend to develop the disease at a younger age.

The importance of the ApoE gene extends beyond Alzheimer's to how well an individual can recover from any type of brain injury. Carriers of ApoE4 gene variants who hit their heads in car accidents, for example, have longer comas, more bleeding and bruising in the brain, more postinjury seizures, less success with rehabilitation, and are more likely to suffer permanent damage or to die.

It is not entirely understood how ApoE influences brain recovery, but the gene is involved in the brain's inflammatory response following head trauma, and people who have an ApoE4 variant take longer to clear their brains of a protein called amyloid, which floods in when the brain is injured. Several studies have found that athletes with ApoE4 variants who get hit in the head take longer to recover and are at greater risk of suffering dementia later in life.

A 1997 study determined that boxers with an ApoE4 copy scored worse on tests of brain impairment than boxers with similar length careers who did not have an ApoE4 copy. Three boxers in the study had severe brain function impairment, and all three had an ApoE4 gene variant. In 2000, a study of fifty-three active pro football players concluded that three factors caused certain players to score lower than their peers on tests of brain function: 1) age, 2) having been hit in the head often, and 3) having an ApoE4 variant.

In 2002, at age forty, former Houston Oilers and Miami Dolphins linebacker John Grimsley began to show signs of dementia. His family noticed that he would repeat the same question, that he could not remember what groceries to buy without a list, and that he would ask to rent movies he had already seen.

Though an experienced hunting guide, Grimsley accidentally shot and killed himself in 2008 while cleaning one of his guns. Grimsley's wife, Virginia, had long wondered whether the concussions her hus-band suffered had anything to do with his mental deterioration, so she donated his brain to Boston University's Center for the Study of Traumatic Encephalopathy.

It was the first of many brains belonging to former NFL players that the BU researchers would examine en route to increasing awareness of the danger of brain trauma in sports. The researchers at the center found an extensive buildup of protein in Grimsley's brain, characteris-tic of chronic traumatic encephalopathy, or CTE. The condition has now been found in scores of brains from college and pro football play-ers. The BU scientists also found that Grimsley—like just 2 percent of the population—had two copies of the ApoE4 gene variant.

In 2009, the BU researchers made national headlines (and head-aches for the NFL) when they reported on dozens of cases of brain damage in boxers and football players. What went entirely unmen-tioned in media coverage, though, was that five of nine brain-damaged boxers and football players who had genetic data included in the re-port had an ApoE4 variant. That's 56 percent, between double and triple the proportion in the general population. Brandon Colby, a Los Angeles–based physician who treats former NFL players says of those patients: "Of the ones who have noticeable issues from head trauma, every single one had an ApoE4 copy." Colby now offers ApoE testing of children to parents who want to weigh the risks of playing football.

Neurologist Barry Jordan, coauthor of the 2000 study of fifty-three football players, and former chief medical officer of the New York State Athletic Commission, once considered making genetic screen-ing for the ApoE4 variant mandatory for all boxers in New York. "I don't think you can stop an athlete from participating," Jordan says, "but it might help just in monitoring them closely. [An ApoE4 gene variant] doesn't seem to increase the risk of concussion, and I wouldn't expect it to, but it may affect your recovery following concussion."

Ultimately, Jordan decided not to implement mandatory genetic testing, primarily because he was concerned about how the information could be used. "Even with [the Genetic Information Nondiscrimination Act]," Jordan says, "you never know. Information still gets out. I think genetic testing is something you can educate athletes about. But I'm not sure how interested people would be in it. Some people don't want to know." Or, as James P. Kelly, a neurologist who was on the Colorado State Boxing Commission, put it: "With ApoE4, some would argue that knowledge is not power."

It's fraught territory, but most current or former pro athletes to whom I explained an ApoE4 test seemed eager to take one, provided the information would be kept from teams, insurance companies, and potential future employers.* Weeks after his visit to Dr. Braverman, Ron Duguay learned that he did indeed have an ApoE4 variant. Had he known of this potential additional risk factor for cognitive impairment, Duguay says he "would've seriously considered wearing a helmet" during his playing days.

Among other athletes I asked about their interest in ApoE testing was Glen Johnson, a professional boxer with seventy-one fights, including wins in 2004 over Roy Jones Jr. and Antonio Tarver. Johnson knew that getting hit in the head—and not simply a particular gene—was the primary factor for brain damage, but says, "I'd never hide from extra information."

Former New England Patriots linebacker Ted Johnson, who suffered a series of concussions that led him to retire and later suffered from amphetamine addiction, depression, memory problems, and chronic headaches, says: "I would be the first person signed up for a test. I wouldn't even hesitate. I know it's no guarantee just because you have this gene, but if it's true that you are potentially at greater risk than the average person, I would do it in a heartbeat. When I was playing we had no information. . . . This kind of information would be

---

*There were exceptions in my interviews, like ex–NFL quarterback Sean Salisbury: "I don't want to know what I'm going to get when I'm eighty-two."

incredible to have if you're a current player." One Alzheimer's researcher at Mount Sinai Hospital in New York has noted that the dementia risk of having a single ApoE4 copy is roughly similar to the risk from playing in the NFL, and that the two together are even more dangerous.

But because the precise degree of additional risk is impossible to quantify, doctors I spoke with almost uniformly felt that ApoE testing should not be offered to athletes. "This is a very controversial area," says Robert C. Green, a BU neurologist who collaborated on the RE-VEAL Study, which examined how people who volunteer for ApoE screening react when they get bad news. "The world of genetics for decades has suggested that there's no reason to give people genetic-risk information unless there's something proven you can do about it." REVEAL found, though, that people who learned they had an ApoE4 variant did not experience undue dread. Rather, study subjects who got bad news tended to increase healthy lifestyle habits like exercise, which doctors told them might help, even though there is no proven remedy for delaying the onset of Alzheimer's.

Still, the doctors' hesitation is understandable. "If we have a gene we know increases your risk of blowing out your knee, if that got into the wrong hands, somebody could decide not to sign a player," says Barry Jordan, the former New York athletic commission medical officer. "That would be a potential problem." (Of course, teams already go to great lengths to guess that same information using physical examinations and medical histories.)

Actually, genes *have* been identified that appear to alter one's risk of blowing out a knee. Biologists at South Africa's University of Cape Town have been leading the way in identifying genes that predispose exercisers to injuring tendons and ligaments. The researchers focused on genes like COL1A1 and COL5A1 that code for the proteins that make up collagen fibrils, the basic building blocks of tendons, ligaments, and skin. Collagen is sometimes referred to as the body's glue, holding connective tissues in proper form.

People with a certain mutation in the COL1A1 gene have brittle bone disease and suffer fractures easily. A particular mutation in the COL5A1 gene causes Ehlers-Danlos syndrome, which confers hyperflexibility. "Those people in the old days of the circus who used to fold themselves into a box, I bet you in most cases they had Ehlers-Danlos syndrome," says Malcolm Collins, one of the Cape Town biologists and a leader in the study of collagen genes. "They could twist their bodies into positions that you and I can't because they've got very abnormal collagen fibrils."

Ehlers-Danlos syndrome is rare, but Collins and colleagues have demonstrated that much more common variations in collagen genes influence both flexibility and an individual's risk of injuries to the connective tissues, like Achilles tendon rupture.* Using that research, the company Gknowmix offers collagen gene tests that doctors can order for patients.

"All we can say to an athlete with a particular genetic profile is that you are at increased risk of injury based on our current knowledge," Collins says. "It's no different than saying that smoking a cigarette increases your chance of lung cancer. The difference is that you can stop smoking, but you can't change your DNA. But there are other factors which you can change. You can modify whatever training you're doing to reduce risk, or you can do 'prehabilitation' training to strengthen the area that is at risk."

A gaggle of NFL players have already availed themselves of testing for "injury genes" that may predispose them to Achilles tendon injuries or torn ACLs in the knee. Duke University's football team, as just

---

*Work on the COL5A1 gene has also found that people with a particular variant are less flexible and may have a benefit in running. The connection could be stiffness of the Achilles tendon, which would allow it to store more elastic energy—again, recall high jump champion Stefan Holm and his stiff Achilles tendon—and improve running economy. In one novel study, athletes with the "inflexible" version of the gene were faster in the running section of an Ironman, but not in the swim or the bike. That is, only in the part of the race when they fully engaged their Achilles tendon did they perform better. The inflexible gene variant, though, is also associated with increased risk of Achilles tendon injury.

one example, sought university approval to submit players' DNA to a researcher on campus who would look for genes that predispose players to tendon and ligament injuries.

So specific genes have now been implicated in sudden death, brain damage, and injury on the field. And now researchers have begun to identify genes that undergird another unpleasant and unavoidable aspects of sports: pain. Genes, it seems, influence our very perception of it.

In the waning years of a career that spanned thirteen NFL seasons, 3,479 carries, a bevy of broken ribs, several separated shoulders, a couple of concussions, a torn groin muscle, a bruised sternum, and a legion of knee and ankle surgeries, 255-pound running back Jerome Bettis developed a Monday-morning tradition. He would sit at the top of his staircase and scoot down toward breakfast on his butt, one step at a time.

On Sundays, the Steelers expected Bettis to run *through* people. "That was my skill set," he says. "It wasn't like I could run away from them." In one game against the Jacksonville Jaguars, a defensive player's thumb came through Bettis's face mask and broke his nose. Team doctors taped the nose and stuffed it with cotton. That helped, until a head-on collision late in the game propelled the cotton up through his nasal passage, down his throat, and into his stomach. "It was like, 'Guys, wait a second, the padding is gone,'" Bettis says. "It was the worst."

No wonder Bettis was unable to walk down the stairs on Monday mornings. The pain was so intense at times that he figured he would have to miss the next game. But once he stepped on the turf Sunday, he never backed down. "When you get on the field, it's not even a question mark," Bettis says. "You do your job, by any means necessary."

Bettis was renowned for his toughness, but he says there are athletes, even in the NFL, who struggle to manage discomfort. "I think some people's bodies kind of shut down from the pain, and it doesn't allow them to still have peak performances," Bettis says. "I saw that problem at times."

Pain tolerance and pain management are as central to most high-level sports as running and jumping, and just why some people tolerate pain better than others is a topic of research at the Pain Genetics Lab at McGill University in Montreal. One room in the lab is stacked from floor to ceiling with clear tanks that house mice, all bred for the study of genes that influence how they (and humans) experience pain, and how that pain can be ameliorated.

In one tank are mice missing oxytocin receptors. They are used in the study of pain, but the mice also have deficits in social recognition. Put them with mice they grew up with and they won't recognize them. In another corner is a tank of raven-haired mice that were bred to be prone to head pain, that is, migraines. They spend a lot of time scratching their foreheads and shuddering, and they are apparently justified in using the old headache excuse to avoid mating. "This experiment has taken years," says Jeffrey Mogil, head of the lab, of the work that seeks to help develop migraine treatments, "because they breed really, really badly."

On another shelf is a tank of mice with nonfunctioning versions of the melanocortin 1 receptor gene, or MC1R. In plain language, they're redheads. It's the same gene mutation that is responsible for the ginger locks of most human redheads. Mogil found that both people and rodents with the redhead mutation have higher tolerance for certain types of pain, and require less morphine for relief.

MC1R was among the first genes identified that influence how humans experience pain. Another was discovered by scientists who followed the theatrical talents of a ten-year-old Pakistani street performer.

Medical workers in Lahore knew the boy well, because after he stuck knives through his arms and stood on burning coals he would come in to get stitched back together. But they never treated him for pain. The boy could feel no pain.

By the time British geneticists traveled to Pakistan to study him, the boy had died, at the age of fourteen, after jumping off a roof to impress his friends. But the scientists found the same condition in six of the boy's extended relatives. "None knew what pain felt like," the scientists wrote, "although the older individuals realized what actions should elicit pain (including acting as if in pain after football tackles)."

The "older individuals" were just ten, twelve, and fourteen. People born with congenital insensitivity to pain tend not to live very long. They don't shift their weight when sitting, sleeping, or standing as the rest of us do instinctively, and they die from the joint infections that result.

Each of the Pakistani relatives with pain immunity had a very rare mutation in the SCN9A gene. The mutation blocked pain signals that normally travel from nerves to the brain. A different mutation in SCN9A causes those who carry it to be hypersensitive to pain, bothered by warmth so easily that they won't wear shoes. In 2010, the British geneticists teamed up with researchers in the United States, Finland, and the Netherlands for a study that reported that much more common variations in SCN9A influence how sensitive adults are to common types of pain, like back trouble. Genetic variation among individuals, it seems, ensures that none of us can truly *know* another's physical pain.

The gene that has been most studied for its involvement in pain modulation is the COMT gene, which is involved in the metabolism of neurotransmitters in the brain, including dopamine. Two common versions of COMT are known as "Val" and "Met," based on whether a specific part of the gene's DNA sequence codes for the amino acid valine or methionine.

In both mice and humans, the Met version is less effective at clearing dopamine, which leaves higher levels in the frontal cortex. Cognitive testing and brain imaging studies have found that subjects with two Met versions—both animals and humans—tend to do better on and require less metabolic effort for cognitive and memory tasks, but that they are also more prone to anxiety and more sensitive to pain. (Anxiety, or "catastrophizing," is a strong predictor of an individual's pain sensitivity.) Conversely, Val/Val carriers seem to do slightly worse on cognitive tests that require rapid mental flexibility, but may be more resilient to stress and pain. (They also get a better boost from Ritalin, which increases dopamine in the frontal cortex.) Additionally, COMT is involved in metabolism of norepinephrine, which is released in response to stress and has a protective effect.

David Goldman, chief of the Laboratory of Neurogenetics at the NIH's National Institute on Alcohol Abuse and Alcoholism, coined the phrase "warrior/worrier gene" to describe the apparent tradeoffs of the two COMT variants. Both versions are common everywhere in the world they've been studied. In the United States, Goldman says, 16 percent of people are Met/Met; 48 percent are Met/Val; and 36 percent are Val/Val, leading him to suggest that both warriors and worriers are needed in every society, so there is widespread preservation of both forms of the gene. "We've never done the study," Goldman says, "but I predict if I took a big group of NFL linemen that they would tend to have the Val genotype, because they're in the trenches every day and they're exposed to pain and they just have to have this super resilience and toughness." *

In fairness, studies of the COMT gene have often been contradictory, and the gene's relevance to pain sensitivity is hotly debated among pain researchers. But the idea that genes involved in emotional regulation might alter pain sensation is uncontroversial. Morphine, after all, doesn't so much decrease pain intensity, but rather reduces the emotional unpleasantness that results from pain. "The pain circuitry is shared so strongly with the circuitry of emotion," Goldman says, "and

---

*The increase in dopamine in the frontal cortex might be helpful for baseball hitters who need to be "hyperalert" and mentally flexible, Goldman says. Amphetamines raise dopamine levels and were a staple of baseball for decades, where they were known colloquially as "greenies." MLB banned amphetamines in 2006, and suddenly the number of players who received doctors' prescriptions for ADHD drugs, stimulants similar to amphetamine, jumped from 28 to 103 in a single season. One doctor I interviewed who works with major leaguers said that he prescribed Adderall to eight pro players who came in with ADHD symptoms. "The diagnostic is an interview," the doctor says, "and it can be very easy to fake." All eight players, he says, had higher batting averages the following season.

many of the neurotransmitters are too. As you modify emotion, you strongly modify pain response."

And sports can be strong modifiers.

Haverford College psychologist Wendy Sternberg was giving a lecture on stress-induced analgesia—the brain's ability to block pain in high-pressure situations—when a student told her that it sounded just like what happens to athletes in competition.

A 2004 Ultimate Fighting Championship heavyweight title fight is an excruciating example. Brazilian jiu-jitsu black belt Frank Mir caught 6'8" Tim "The Maine-iac" Sylvia in a joint lock called an arm-bar. Mir grabbed Sylvia's extended right arm, braced the elbow joint against his hip, and pulled backward so forcefully it looked as if he were heaving back a train brake.

The pop of Sylvia's shattering arm was audible on television. Referee Herb Dean rushed in to separate the fighters and shouted for the match to stop. Sylvia set to cursing and demanding that the fight continue. Only later, as he sat on a gurney en route to the hospital, did Sylvia begin to feel pain and realize that his attempt to keep fighting had been ill considered. It took three titanium plates to rig his arm back together. "[The ref] probably saved my career," Sylvia says, because in the heat of battle he couldn't perceive the pain on his own.

Says Sternberg, "Under conditions of acute stress the brain inhibits pain, so you can fight or flee without worrying about a broken bone." A system to block pain in extreme situations evolved in the genes of all humans, and even quotidian sports settings tap into it.

In 1998, prompted by her student's suggestion, Sternberg tested the sensitivity of Haverford track athletes, fencers, and basketball players to cold and heat pain two days before they competed, on the day of a competition, and two days later. She found that basketball players and runners were less sensitive to pain than their nonathlete peers to begin with, and that all of the athletes were least sensitive to pain on game day. "I think athletic competition can activate the fight-or-flight mechanism," Sternberg says. "When you get in a competition that you care about, you're going to activate it."

Pain can be modified by a game situation or by the emotions of an athlete, but the genetic blueprint for pain in the body is encoded in the brain, whether or not that body even exists in its entirety. (People who are born without limbs or who have them amputated nonetheless often experience pain in those "phantom limbs.") Still, pain must be *practiced* in the first place.

In the 1950s, Canadian psychologist Ronald Melzack was working toward his Ph.D. at McGill under psychologist D. O. Hebb, who was studying how extreme deprivation of life experience affects intellect. Hebb was experimenting on Scottish terriers.

The dogs were well cared for, groomed, and fed, but they were totally isolated from the outside world. Hebb was interested in how that would alter their ability to navigate a maze. (The answer: very negatively.) But it was in the holding room, before the maze, where Melzack made the observation that started him down the road to becoming the most influential pain researcher in the world. "The water pipes in the room were at head level for the dogs," Melzack says, "and these wonderful dogs would run around and bang their heads right into the pipes, as if they felt nothing. And they kept running around and banging their heads on the water pipes."

Melzack was a smoker at the time, so he struck a match. "I held it out, and they'd put their nose in it," he says. They'd back up, "and then come back and sniff it again. I'd put it out and light another match, and they'd sniff it again and again." The dogs obviously had normal cerebral hardware, but had missed the critical developmental window for downloading the brain's pain software. They never learned to be deterred by the flame. Just like language, or hitting a baseball, even though each of us may be born with the requisite genetic hardware, if we miss the window for acquiring the software, the genes are of little use. Adds Jeffrey Mogil, of McGill's Pain Genetics Lab: "The fact that something like pain would have to be learned at all is pretty surprising."

Pain is innate, but it also must be learned. It is unavoidable, and yet modifiable. It is common to all people and all athletes but never experienced quite the same way by any two individuals or even by the same individual in two different situations. Each of us is like the hero in a Greek tragedy, circumscribed by nature, but left to alter our fate within the boundaries. "Maybe if you're a worrier by genotype, it's a better idea not to be a warrior by profession," says Goldman, the neurogeneticist. "Then again, it's hard to say, because people overcome so much."

Like most traits discussed in this book, an athlete's ability to deal with pain is a braid of nature and nurture so intricately and thoroughly intertwined as to become a single vine. As one scientist told me: without both genes and environments, there are no outcomes.

It reinforces the idea that any notion of finding an "athlete gene" was a figment of the era of wishful thinking that crested a decade ago with the first full sequencing of the human genome, before scientists understood how much they don't understand about the complexity of the genetic recipe book. What, exactly, most human genes do is still a mystery. Sure, the ACTN3 gene may tell a billion or so people in the world that they won't be in the Olympic 100-meter final, but chances are they all already knew that.

If thousands of DNA variations are needed to explain just a portion of the differences in height among people, what are the chances of ever finding a single gene that makes a star athlete? Slim? Or none?

And yet . . .

16

The Gold Medal Mutation

t is December 2010, and human civilization in northern Scandinavia is temporarily reduced to a layer of sediment beneath the snow. Excavation will come only with spring. The last few days have seen record snow and a constant -15 degrees Fahrenheit at the Arctic Circle in Finland—the Napapiiri, as the Finns call it—where I am now. There's no wind, so the first crunching step outside each morning is deceptively placid, before nose hairs morph into ice daggers.

The Finns call this part of the year "*Kaamos* time." There is no exact English word for *Kaamos*, but it roughly means polar night. It means the time of the year when northern Finland is tilted so far from the sun that daylight is really three hours of twilight that around two P.M. flickers out as if under a cosmic candle snuffer.

I'm driving north along highway E8, in search of a ghost. And this is the perfect place for one to live—among the pines and spruce made hard by the cold and made white by the snow; beside the Swedish whitebeam and the European white elm; and amid the silver birch and the downy birch with their white skins wrapped in a blanket of white mist. Reindeer prance beside the road and disappear into whorls of snow. It is all thick and white, as if some celestial milk bottle had toppled over and I'm driving through the puddle. This is a land of austere beauty, of the most gleaming whites of sky and snow and the most vacuous blacks of night.

But Iiris Mäntyranta was born not far from here, and she can see colors. To her, the sky has a bluish tinge, and the walls of cloud give passage to the occasional spangle of purple light.

Before I made contact with Iiris months earlier, I was not sure whether my ghost—her father—was even still alive. His words had not appeared in any English-language press I could track down since the 1960s, when he emerged from his tiny Arctic hamlet and won seven Olympic medals, three of them gold. Now we are traveling north, together, to meet him.

After three hours of driving from Luleå, Sweden, where Iiris works as a county government administrator, we are getting close. Just past the Arctic Circle, we drive through Pello, a town of four thousand that is the last semblance of a city we will see along the road. On our way out of Pello, we pass a granite pedestal atop which sits a larger-than-life bronze statue of a man in mid-cross-country ski stride. The man is Iiris's father.

A half hour later we pull off the paved road and drive down a narrow pass among the pine trees. We stop in front of a cream-colored house on the west side of a large lake. As I get out of the car, I'm conscious of being watched. I turn to look back at the pass down which we came. A sandy-colored reindeer has come around the corner and has its gaze fixed on me, as if it can smell the Brooklyn on my clothes. It's frigid and snowing, so we hurry inside the house.

No sooner do I step inside and kick the frost from my boots onto the welcome mat below the rifle rack than an oddly Mediterranean face appears in the entryway. It is the man from the statue, the great Eero Mäntyranta. I'm taken aback. In pictures I had seen of him from the 1960s his skin was perhaps slightly too dark for the Arctic, but it was nothing that would warrant a second look. But now he is closer to the hue of the red paint that comes from this region's iron-rich soil than to that of the snow. Iiris told me on the drive up that her father's unique gene mutation had caused his skin to redden as he got older, but I didn't quite expect this shade of cardinal, mottled in places with purple.

The contrast is stark when Eero's wife, Rakel, with her glacier blue eyes and alabaster skin, steps into the entryway. Eero speaks no English, but he greets me with a wide smile. Everything about him has a certain width to it. The bulbous nose in the middle of a softly rounded face. His thick fingers, broad jaw, and a barrel chest covered by a red knit sweater with stern-faced reindeer across the middle. He is a remarkable-looking man. His dark hair is meticulously slicked back and he has prominent cheekbones that seem to draw up the edges of his thin lips so that he looks constantly pleased and inquisitive. There is also an unmistakable strength about him, never mind that he is seventy-three years old. The middle finger of his right hand is bent sharply at the top joint, a periscope peering at the index finger. His hands look as if they could snap a ski pole in two, a supposition backed up by his handshake.

Eero ushers me to the kitchen where Rakel serves tea and coffee to me and to Iiris and to Iiris's Swedish husband, Tommy, as well as to Iiris's son Viktor, a musician—his band Surunmaa plays a fusion of folk, blues, and tango—who is staying in a cabin on Eero's property while he films a documentary about his grandfather's life.

The wide windows in the kitchen look out into the snowy forest. This used to be an area of extreme poverty, but now even the remote north of Finland has prospered from the country's trade in timber and electronics, and the residences are as impeccably kept as dollhouses. Sitting here sipping tea from a tiny porcelain cup, grinning at a red-nosed man in a reindeer sweater, I feel certain that I have stepped into a Christmas snow globe.

After introductions and tea, I follow Eero outside where he feeds a dozen reindeer handfuls of pale green lichen. The reindeer are used for racing and also for meat. When I walk up to one of the animals, Viktor translates Eero's warning that, unlike horses, reindeer do not like to be touched by human hands. Some of the reindeer are teddy bear brown, and others are chalk white. Outside, against the falling snow, the redness of Eero's face is in its greatest relief.

With daylight quickly fading, we return indoors. Over the next few hours I interrogate Eero about his remarkable athletic career. Iiris,

Tommy, and Viktor take turns translating the language that to my ears sounds like a stream of deep "ess's," punctuated by crackling "k's" and "cox's" spliced with the occasional Spanish-sounding rolling "r."

When the sun fades, we will take a break from talking for a meal of reindeer meat and potatoes. And Eero will laugh deeply when the fork he is holding returns his mind to a time more than forty years ago, when he was one of the greatest athletes in the world.

It was 1964 and Eero Mäntyranta was once again in the uncomfortable position of honored guest. Surrounded by the clinking of crystal, he furrowed his heavy eyebrows at the three forks flanking his plate. He had just won two golds and a silver medal at the Winter Olympics in Innsbruck, Austria, dominating the cross-country skiing competition to such an extent that the media deemed him "Mr. Seefeld," a reference to the competition venue. In the 15K race, Mäntyranta finished forty seconds ahead of the next skier—a margin of victory never equaled in that event at the Olympics before or since—while the next five finishers were within twenty seconds of one another. In the 30K race he won by over a minute. Now came the hard part: dinner. Becoming one of your nation's all-time great athletes necessarily comes with a glut of honorary feasts.

After his first gold, in a relay at the 1960 Games in Squaw Valley, California, Mäntyranta attended a celebratory meal in Los Angeles organized by Finland's Olympic committee. That time, he was about to drink from a goblet on the table when a group of urbane guests strode up and began washing their hands in it. But the three forks presented a new puzzle.

When Mäntyranta was a child growing up in rural Lankojärvi, Finland, in the 1940s, his family shared a single fork. It was passed around the 170-square-foot room that was their house, overlooking the lake for which the town was named. In lieu of cutlery, the children used sharpened sticks to spear chunks of potato and slices of bread.

The Mäntyranta brood would have numbered twelve had all the

children survived. As it was, they were six. Still, with Eero, his parents, his brothers and sisters, and his older sister's husband, the single room could be a bit cozy. Add the neighbors who would stop by to shoot the breeze and have a cigarette, and it was not uncommon to have a dozen people in the room. In that atmosphere, young Eero first employed the admirable capacity for solitary focus that would later serve him well during the lonely training hours on the ski trail under the black sky of *Kaamos* time. He was an excellent student, only because he could block out the commotion in the room, curl up underneath the smoke, and do his schoolwork by the fickle light of an oil lamp. Those were spare days in postbellum Finland, with the country locked in two decades of war debt to the Soviet Union.

Eero was only six years old in the winter of 1943, when Nazi soldiers pushed north and Lankojärvi was evacuated. He was put on a truck with all the women and children from the town and told by a Finnish soldier to keep quiet lest German soldiers hear them. He shuddered when one old woman refused to heed the advice and belted out Communist work songs. The truck eventually made it to a ferry that took them across the border into Overtornea, Sweden, where Eero gazed in wonderment at the bullet shells that lay across the ground like a dusting of leaden snow. He and his family stayed in Sundsvall, Sweden, through the winter, until they were allowed to return to a Finland free of snow and free of Nazis.

The trek back home in the spring was a journey of diminishing hope. They had to take a horse and carriage through the woods because the roads were strewn with live landmines. The German military set fires on their way out of Finland, and in a country whose towns are but clearings amid dense forest, tinder was in hearty supply. Lapland burned like a vast fire pit, making smoldering embers of what were once doorjambs, staircases, and gables crafted from pine.

But the Mäntyrantas returned to find their home one of the few standing. They lived on the remote side of the water, with no road, so Nazi soldiers didn't bother to venture over the lake and through the woods to raze the few nondescript shacks on the other side. The lake had saved their home. The same lake that started Eero's skiing career.

While the Germans didn't try to cross the lake, many of the children of Lankojärvi had no choice. School was on the other shore. Almost as soon as Mäntyranta could walk, he could ski, and within a year of returning from Sweden, he was joining other kids in skating—he once fell through the ice and nearly drowned—or skiing across the lake to school, on nothing more than wooden planks nailed together. It took about an hour to make the trip, and during winter it was pitch black the entire way, so the kids would simply aim at the far shore and hope for the best.

Out of necessity, everyone in Lapland skied. But it did not take long for Mäntyranta to stand out. As early as seven years old he would win the cross-country-ski races at school. When he was ten, he started winning the races that brought kids from local villages together. At eleven, he polished off the youth competition in the entire municipality of Pello.

Unlike the Finnish youth in the south, Mäntyranta never dreamed about sporting glory as a boy. Sports had been integral to Finland's identity ever since the country declared independence from Russia in 1917. National sports organizations were formed, and they paid off in spades—and medals. The "Flying Finn" distance runners dominated the world in the 1920s. After World War II, when Helsinki was awarded the 1952 Olympics, sport again became a beacon of unity for the Finnish people. But Finland's sporting tradition had no impact on young Eero. With no radio or newspapers in Lankojärvi, he had no idea who the great Finnish athletes were. He didn't have the chance to be inspired by the words of the beloved Finnish runner Paavo Nurmi, who told the world, "Mind is everything. Muscle—pieces of rubber. All that I am, I am because of my mind." Mäntyranta's only exposure to the '52 Helsinki Games was a picture that he saw in a neighbor's house of a Brazilian man triple-jumping. For Eero Mäntyranta, skiing was a mode of transportation and a chance for a better job.

For twenty years after the end of the war, Finland's economic growth was stunted by having to send surplus money and resources to

Russia to pay off war debt, so the only job for a young man in Lapland was cutting and hauling wood from the forest. At fifteen, Mäntyranta was living in the forest among grown men, many of them criminals who came to the far north to evade the law. The men spent their leisure time drinking, playing cards, and fighting. Mäntyranta slept with a block of wood under his pillow in case he had to bludgeon an attacker during the night. It was both a harrowing and exciting existence for a young man. But after two years, he'd had enough.

He knew that the government had a habit of giving promising young cross-country skiers cushy jobs as border patrol guards where they could essentially ski along the border for both training and work. So he started training in his free time from forest work, and his progress was stunning. At nineteen, he traveled to Switzerland for a series of races that, if he performed well, would push him toward the Finnish national team. He won all of them, and a job as a border patrol guard followed soon thereafter.

Mäntyranta's mother warned him that it was time to save money, not to go chasing girls. He heeded that advice for a good two weeks, until he spent a night in Pello dancing with his blond-haired, blue-eyed future wife. When the couple later had children, Mäntyranta would often train in the summer by sending Rakel and the kids off in a car to their cottage twenty miles away. He would then run or walk to meet them.

Despite regular smuggling over the Sweden/Finland boundary, the border north of the Arctic Circle was generally quiet, especially in winter, so Mäntyranta had plenty of time to throw himself into training. At 5'7" (in thick socks), he was very small for a cross-country skier. With black eyebrows arcing over dark brown eyes, and a slight tan complexion to his skin, he looked more like someone born of an Italian beach, not the lower Arctic pine forest. But there he was, for fifty miles a day, jabbing away with his poles at the snowy blanket that covered the earth. He often trained by the moonlight. Or, if he was near the road in Pello, he would exist for a moment in the beams of a passing car, before fading back into the darkness. When the moon was obscured, he worried that he would ski headlong into trees, but he managed to stay clear of accidents, and his progress continued at a remarkable clip.

By twenty-two years old, he was clearly good enough to ski for Finland at the 1960 Olympics, but most of the best skiers were older, and team officials weren't eager to allow an inexperienced skier to test his mettle on the biggest stage. Mäntyranta persuaded the team managers to allow an intramural time trial. He placed second, behind thirty-five-year-old skiing legend Veikko Hakulinen, who already had two Olympic golds. The performance earned Mäntyranta a spot on the 4×10K relay team at the Games, where they took home the gold.

That Olympic title was just the preamble. Two golds and a silver followed in Innsbruck in 1964. Then a silver and two bronzes in Grenoble, France, in '68, and a bevy of world championship medals along the way. In all, he placed in five hundred races, amassing enough crystal glasses and silver bowls and dishes to fill a china shop. Even now, he will awake some days and tell Rakel that his legs are tired because he was again ski racing in his dreams.

But Mäntyranta's trail to the skiing pantheon started long before the 1960 Games. It started before the work in the forest prodded him to seek a better life. Before he began skiing across the lake to school on warped planks. Even before he first stood on skis when he was three years old. It began when his great-grandfather made the trip to Finland.

The details of the Mäntyranta family's beginnings in Finland are murky, but relatives were certainly in Lapland by the 1850s. It was probably Eero's great-grandfather who came from Belgium to work as a blacksmith, forging coins. His son, Isak, married a woman named Johanna, whose father was just wealthy enough to own a swath of land north of Lankojärvi. Isak and Johanna lived in a cottage on the land on the condition that Isak help the resident farmers there with their work. But Isak was not one for manual labor and he soon wore out his welcome.

Eero would not inherit Isak's lax work ethic, but—via his father,

Juho—he would inherit a rare version of a gene that altered his body's blood supply.

The first sign in Eero was during a routine medical exam when he was a teenager. A blood test showed that he had extraordinarily high levels of hemoglobin, the oxygen-carrying protein in red blood cells. It is the iron in hemoglobin that gives blood its red color. Because Eero was perfectly healthy, there was little concern about his high hemoglobin levels.

But that began to change during his competitive career. Every time he was examined, Eero was found to have high hemoglobin and far more than the usual amount of red blood cells. Normally, those are signs that an endurance athlete is blood doping, often with a synthetic version of the hormone erythropoietin, or EPO. EPO signals the body to produce red blood cells, so injecting it spurs an athlete's own body to bolster its blood supply.

At times, Eero's extraordinary red blood cell count—measured at up to 65 percent higher than that of an average man—sullied his sterling career. Despite the fact that his blood levels had been documented since he was a kid, speculation was rife that his unusual blood profile was the result of doping. It was not until twenty years after his retirement from skiing that scientists pinpointed the truth.

From time to time, other members of the Mäntyranta family would discover through a routine medical test that they had elevated hemoglobin levels. Because there were no apparent ill health effects, doctors did nothing about it.

It was enough, however, to ignite the curiosity of Pekka Vuopio, the head of hematology at the University of Helsinki and a native Laplander who knew well the athletic exploits of Eero Mäntyranta. In 1990, Vuopio and his colleagues invited Eero to Helsinki for a series of tests in the hope that examining him might shed light on a condition called polycythemia, an elevation in red cells that can cause a dangerous thickening of the blood and that sometimes runs in families.

One of the doctors' first theories was that Eero's red blood cells might have a longer life span than normal, so that new blood cells were produced before old ones had been cleared away. But that turned out not to be the answer. Another possibility was that Eero naturally secreted high levels of EPO, thus instructing his body to overproduce red blood cells. But that wasn't it either. The level of EPO in Eero's blood was so low that it was nearly below the lower limit for healthy adult men.

But when hematologist Eeva Juvonen examined Eero's bone marrow cells in the lab, she saw something astonishing. In order to test whether his bone marrow cells—which produce red blood cells— were particularly sensitive to EPO, the research protocol was to add EPO to a cell sample and track red blood cell production. Eero's bone marrow cells began the process of creating red blood cells before Juvonen could even stimulate them with EPO. Whatever tiny speck of EPO that was already in the sample was enough to keep the red cell factories humming. So it was clear that Eero's body heeded the call of even trace quantities of EPO with extraordinary vigor. Illuminating the reason why would require more members of the Mäntyranta clan.

Albert de la Chapelle identifies himself as a gene hunter. He is exceedingly good at tracking his prey. He is the geneticist who argued on behalf of María José Martínez-Patiño when she was barred from competing as a woman. These days he spends his time at Ohio State University training his sights on the genes that predispose people to the most deadly cancers ever known, like acute myeloid leukemia, which interferes with blood cell production and can put a previously healthy patient in the ground in a matter of weeks.

De la Chapelle spent most of his career at the University of Helsinki, hunting gene mutations that cause diseases that show up in Finland far more often than in the rest of the world. These diseases come from so-called founder mutations, meaning that a mutation arose in a member of a small group and spread through that population as it grew. De la Chapelle was part of a team that clarified the genetic basis of more than twenty diseases—multiple forms of epilepsies and dwarfisms among them—that are endemic to Finland. (And sometimes to Minnesota, a state heavy with residents of Finnish ancestry.)

Not long after Eero Mäntyranta's blood was examined in the lab, de la Chapelle made a trip to Lankojärvi to meet a group of forty Mäntyrantas who had assembled at Eero's house to talk with the researchers who were now studying their blood. It was winter, and de la Chapelle remembers marveling at the noontime sun as it kissed the surface of the lake.

After a lunch of fresh reindeer prepared by Rakel, de la Chapelle set to mingling in the living room. "I was sitting there on the couch with these three elderly ladies," de la Chapelle recalls, "and I already knew two had the condition and one did not. And they went over their health with me and it was the one without the condition that had all the health problems and the two with it were quite healthy and were unaware of anything at all being different with them."

Even if not for their slightly darker complexions, de la Chapelle would have known that the two healthy women had the blood condition. He had already been through their genomes.

In all, ninety-seven Mäntyrantas were examined, twenty-nine of whom had remarkably high hemoglobin, along with slightly ruddier complexions than the average Finn. Unlike the initial study of Eero, this examination went more than blood deep. De la Chapelle probed all the way down to a particular gene on the nineteenth chromosome, the EPOR, or erythropoietin receptor gene.

This particular gene tells the body how to build the EPO receptor, a molecule that sits atop bone marrow cells awaiting the EPO hormone. If the EPO receptor is a keyhole, it is one made specifically to accept only the key that is the EPO hormone. Once the key is in the lock, the production of red blood cells proceeds. The receptor signals a bone marrow cell to start the process of creating a red blood cell that contains hemoglobin.

Of the 7,138 pairs of bases that make up the EPO receptor gene, there was a single base that was different in the twenty-nine family members who had unusually elevated hemoglobin levels. Each family member, like every human being, had two copies of the EPOR gene. But at position 6,002 in only one copy of each affected family member's two EPOR genes, there was an adenine molecule instead of a guanine molecule. A minuscule alteration, but the impact was immense.

Instead of adding information for the cellular machinery to continue to build the EPO receptor, the spelling change constituted a "stop codon," the genetic equivalent of a period at the end of the last sentence of a chapter. A stop codon essentially tells RNA—ribonucleic acid, the molecule that reads DNA code so that it can be translated into action—that the instructions are finished. *Move along, nothing more to read here*, it says. So instead of coding for the amino acid tryptophan, as that section of the EPOR gene normally would have, the Mäntyranta family mutation caused the receptor simply to stop being built with over 15 percent of its construction unfinished. The unfinished portion in the affected Mäntyranta family members happens to be a segment of the receptor in the interior of the bone marrow cell. The piece of the receptor on the exterior of the cell awaits the EPO key, while the interior portion modulates the subsequent response, acting like a brake to halt hemoglobin production. In the affected Mäntyrantas, who are missing the brake, the production of red blood cells runs amok.

Fortunately for the family, the overproduction of red blood cells did not lead to ill health. Save for the slightly dark complexion, family members had no outward signs of abnormality and generally discovered their condition by accident during routine checkups.

The Mäntyranta EPOR gene finding was a major discovery in the early 1990s. The high hemoglobin condition in the Mäntyrantas was passed down through the family in an autosomal dominant fashion, meaning that only a single copy of the mutant gene was required for a family member to have the condition. Other dominantly inherited gene mutations had been discovered before that study, but they were generally tied to serious illnesses.

In the papers they published in 1991 and 1993, the researchers noted that Mäntyrantas who carried the family EPOR mutation had long lives. They had found, it seemed, a mutation beneficial for an athlete and otherwise of little consequence. De la Chapelle says, though, that he could never convince Eero himself that the EPOR mutation aided him in his Olympic quest. "He kept saying that it was not his bodily strength," de la Chapelle says, "but his determination and psyche."

Since I came all the way from Brooklyn to meet him, Eero is eager to tell me about his visit to New York City after the 1960 Winter Games. "Scary" is how he describes his first impression of the morass of Cadillacs, streetlights, and asphalt.

He has also laid out for me some of his most prized medals, the seven from the Olympics, and a medal of honor that the government normally reserves for military heroes. As they have for polar night, the Finns have an untranslatable word, *sisu*, that roughly means strength of passion, or calm determination in the face of obstacles. The Finnish government determined that Eero was the embodiment of *sisu*.

Iiris, wearing shoulder-length blond hair and black-rimmed glasses, translates a story from her childhood about the aftermath of the 1964 Olympics, when the local electric company paid for Eero to return home in a helicopter. It landed atop the ice covering the lake amid hundreds of revelers who had gathered to celebrate. Iiris was a little girl, and remembers running excitedly toward the helicopter. At first, Eero enjoyed the attention, and it afforded him a job working for the local government teaching physical education to children. But it quickly became a burden.

Through the mid-1960s, reporters would show up unannounced at

Eero's door asking him to "tell me a story, but not what you tell others," Eero says through Iiris's translation. Before competitions, tourists from southern Finland would drop by asking to see medals and to take pictures, requests that Eero and Rakel felt obliged to honor. For Eero, skiing had always been more about winning and getting a better job than an intrinsic love of the activity, so the unwanted attention was enough to push him to retire from ski racing following the 1968 Olympics, at the age of thirty.

At the behest of a Finnish celebrity magazine, he made a brief comeback before the 1972 Winter Olympics in Sapporo, Japan. He had not skied a stride, or exercised at all, for three years, and he was well above his racing weight. The magazine promised to pay Eero's training expenses so that he could take a break from work so long as he gave the publication access to document his comeback. Eero returned to the trails just six months prior to the Olympics but made the team and finished nineteenth in the 30K race in Japan before going back into retirement, this time for good.

Toward the end of my visit, we all take spots on couches and chairs in the living room, flanked by paintings of winter landscapes. Eero points out a series of sepia photographs hanging on the wall. They are of his ancestors. There is swarthy-skinned Isak, in a vest and newsboy cap, reclining on the ground of a forest clearing and enjoying a meal with Johanna, her head wrapped in a light-colored scarf. And above that is a picture of Eero's parents, Juho and Tynne, sitting on wooden chairs in a patch of cleared land with several of their children.

Isak and Juho died before de la Chapelle ever started probing the family genome, but enough Mäntyrantas were tested that he was able to create a genetic family tree and deduce that they had the EPOR mutation. Juho's two brothers, Leevi and Eemil, also carried the mutation.

But it will soon come to an end down Eero's line. His son Harri had it and showed promise as a youth cross-country skier, but Harri died as a young man of an illness that had no relation to the EPOR mutation. Iiris does not have it, and of Eero's remaining two children, fraternal twins Minna and Vesa, only Minna has it, but her only son does not.

When I ask Eero whether he was relieved that the University of Helsinki doctors lifted the suspicion of blood doping from his victories, he says yes, but that he disagrees with the suggestion that the mutation gave him an advantage. Eero's feeling is that the increased viscosity of his red-cell-loaded blood would have hampered his blood circulation, thus balancing any performance benefits. De la Chapelle disagrees staunchly. "It's an advantage, there's no question," he told me, noting that Eero's hemoglobin levels were the highest he has ever seen. "If the blood didn't circulate well, that would be a pretty serious situation and you would know."

In recent years, Eero has had several bouts of pneumonia that his doctors think could be related to his thick blood, so he is now on blood-thinning medication. Iiris adds that the redness of his skin is also a recent development. During his competitive days, Eero showed no ill effects of his EPOR mutation, and other Mäntyrantas with the mutation have remained healthy into old age.

While the extensive scientific documentation of the Mäntyranta family's mutation is unique in sports, there have certainly been other successful athletes with preternaturally high hemoglobin levels. Endurance sports like cross-country skiing and cycling have set up systems whereby an athlete with abnormally high hemoglobin or red blood cell levels can earn a medical exemption to compete if that athlete can prove that his hemoglobin is naturally elevated. A number of athletes have been given such exemptions, and have gone on to great success.

Italian cyclist Damiano Cunego was granted a medical exemption by the International Cycling Union and at twenty-three years of age became the youngest road cyclist ever to be ranked number one in the world. Frode Estil, a Norwegian cross-country skier who was given an exemption by the International Ski Federation, won two golds and one silver medal at the 2002 Winter Olympics in Salt Lake City. Neither of these men had hemoglobin levels as high as Eero's—the normal range for men is 14 to 17 grams of hemoglobin per deciliter of blood, and Eero was high even compared with his own family members, consistently over 20 and as high as 23—but Cunego and Estil nonetheless had elevated levels that they could prove were natural and that were higher than those of their teammates and competitors who trained in similar manners.

Like the naturally fit six from the York University study, there was just something innately different about them.

With the three-hour drive back to Luleå in mind, Iiris tells Eero and Rakel that she will see them soon for Christmas, and tells me that we should hit the road.

As we are getting ready to leave, I suddenly chide myself for nearly forgetting to ask an obvious question. When I was told that the EPOR mutation will not continue down Eero's direct line of descendants, I was disappointed that there would be no way to see whether it might push younger Mäntyrantas to athletic success. But from de la Chapelle's family tree I know that there are extended family members who have the mutation.

"Do Eero's siblings have the mutation?" I ask Iiris.

One of them does, she tells me. His sister Aune, and two of Aune's children have the mutation, her son Pertti and her daughter Elli.

And did they ski? I ask.

They did, she tells me.

And were they any good?

Elli was twice a world junior champion in the 3×5K relay in 1970 and '71. And Pertti, competing at the site of his uncle's most famous triumphs, won an Olympic gold medal in the 4×10K relay in 1976 at the Innsbruck Winter Games. In 1980 he added a bronze at the Lake Placid Games.

No one else in the family races.

The Perfect Athlete

E ero Mäntyranta's life story is a paragon of a 10,000-hours tale.

Mäntyranta grew up in poverty and had to ski across a frozen lake to get to and from school each day. As a young adult, he took up serious skiing as a way to improve his life station—to land a job as a border patrolman and escape the danger and drudgery of forest work. The faintest taste of success was all Mäntyranta needed to embark on the furious training that forged one of the greatest Olympic athletes of a generation. Who would deny his hard work or the lonely suffering he endured on algid winter nights? Swap skis for feet and the Arctic forest for the Rift Valley and Mäntyranta's tale would fit snugly into the narrative template of a Kenyan marathoner.

If not for a batch of curious scientists who were familiar with Mäntyranta's exploits and invited him to their lab twenty years after his retirement, his story might have remained a pure triumph of nurture. But illumined by the light of genetics, Mäntyranta's life tale looks like something entirely different: 100 percent nature and 100 percent nurture.

Obviously, Mäntyranta had rare talent. Just as clearly, he needed to train assiduously to alchemize that talent into Olympic gold. As psychologist Drew Bailey told me: "Without both genes and environments, there are no outcomes." Instances in which a single gene has a dramatic effect, as in Mäntyranta's case, are extremely rare, and finding athleticism genes is extraordinarily complex and difficult. But a present inability to pinpoint most sports genes doesn't mean they don't exist, and scientists will, slowly, find more of them.

One of the concerns held by Yannis Pitsiladis, the scientist who traverses Africa and Jamaica to collect athlete DNA, is that discovering genes that influence athletic performance will detract from the hard work undertaken by athletes if those genes turn out to be more concentrated in one ethnic group or region than another. But we already know that certain ethnic groups have genes that equip them superiorly or inferiorly for particular athletic endeavors. To use Yale geneticist Kenneth Kidd's example, we can agree that Pygmy populations are unlikely to be founts of NBA stars, given that Pygmies tend to have few gene variants that result in tall stature compared with other populations.

Height is clearly an innate advantage in basketball. But does it detract from Michael Jordan's achievements that he had the good fortune to be endowed with genes that contributed to his being taller than Pygmies, and than most other men on earth? If there exists a scientist or sports fan who would denigrate Jordan's hard work and skill because of his obvious gift of height, I didn't meet him in the reporting of this book. In fact, the opposite extreme—ignoring gifts as if they didn't exist—is much more common in the sports sphere.

Consider this title and subtitle of a *Sports Illustrated* story: "The Fire Inside: Bulls center Joakim Noah doesn't have the incandescent talent of his NBA brethren. But he brings to the game an equally powerful gift." The "gift" is Noah's desire to win. Never mind that he is the 6'11" son of a French Open tennis champion and has a wingspan of 7'1¼" and a 37½" vertical jump. If those aren't incandescent athletic endowments, then what, pray tell, are? Noah's lack of talent referenced in the headline—and by Noah himself in the story—would seem to describe the fact that he's a graceless ball handler and mediocre jump shooter. Which, based on the sports science, probably has more to do with the specific work he has put in to develop dribbling and shooting skills than with his hereditary gifts. A more honest headline might read: "The Talent Outside: Joakim Noah has not acquired basketball-specific skills to the extent of his teammates, but he is at the upper extreme of humanity in terms of his physical gifts and therefore can be a good NBA player anyway."

Acknowledging the existence of talent and of genes that influence athletic potential in no way detracts from the work it takes for that talent to be transformed into achievement. The studies undertaken by K. Anders Ericsson—the so-called father of the 10,000-hours "rule"—and his colleagues typically don't address the existence of genetically based talent because their work begins with subjects of high achievement in music or sports. When most of humanity has already been screened out of a study before it begins, the study often has little or nothing to say about the existence or nonexistence of innate talent.

In reality, any case for sports expertise that leans entirely on either nature or nurture is a straw-man argument. If every athlete in the world were an identical sibling to every other athlete, then only environment and practice would determine who made it to the Olympics or the professional ranks. Conversely, if every athlete in the world trained in exactly the same way, only genes would separate their performances on the field. But neither of those scenarios is ever the case.* (The occasional example of same genes/same training tells the expected story. I was standing beside the finish line of the London Olympic 400-meter final when Belgian identical twins and training partners Kevin and Jonathan Borlée, despite running in lanes on the extreme opposite sides of the track, finished 0.02 of a second apart.) Athletes are essentially always distinguished by both their training environments *and* their genes.

In some cases, as with the ability of baseball hitters to react to a pitch, a skill that seems based on superhuman reflexes is largely the

---

*There are, however, a few interesting anecdotes. Identical twin sisters and elite U.S. sprinters Me'Lisa and Mikele Barber train separately. They have 100-meter personal bests that are 0.07 of a second apart.

result of a learned mental database. (Once the database is in place, however, an athlete who possesses outstanding visual hardware can put it to superior use.) In others, as with the ability to respond rapidly to endurance exercise, genes mediate the very improvements that come from hard training. In all likelihood, we overascribe our skills and traits to either innate talent or training, depending on what fits our personal narratives.

Steve Jobs famously said that he had long thought his personality was entirely the result of his life experiences until, as an adult, he met for the first time novelist Mona Simpson, the sister he did not know he had. Jobs marveled at how similar he was to Simpson despite having grown up with a different family. "I used to be way over on the nurture side, but I've swung way over to the nature side," Jobs told the *New York Times* in 1997. "And it's because of Mona and having kids. My daughter is fourteen months old, and it's already pretty clear what her personality is."

As the study of genes matures, we will increasingly find genetic inputs—some large and many trivial—behind the sports stories we tell. But we are unlikely ever to receive complete answers from genetics alone, and not merely because environment and training are always critical factors. Recall that even for height, an easily measurable trait, scientists needed several thousand subjects and hundreds of thousands of spots of DNA code to account for even half of the variance in height between adults. It is increasingly clear that many traits are influenced by the interplay of large numbers of DNA variations. Thus, studies will require hundreds or even thousands of subjects to get at the genetic root of such traits. But there aren't thousands of elite 100-meter runners in the world. Additionally, the gene variants that make one sprinter fast may be completely distinct from those that make her competitor in the next lane fast. Remember, with HCM, the disease that leads to sudden death in athletes, most of the distinct, known gene variants that cause the disease are "private" mutations.

That is, they have thus far been located only in a single family. The same physical outcome can sometimes be reached via many different genetic pathways.

Nonetheless, as I am writing this, headlines are erupting with the news that Japanese scientists have created fertile eggs from mouse stem cells. On the radio, a scientist just speculated that the breakthrough will ultimately lead to the ability to engineer offspring for specific traits, including athleticism. *We can build the perfect athlete*, the scientist implied. "It will give parents a great ability to choose the genetic traits of their children," Stanford bioethicist Hank Greely told NPR.

With respect to athletic traits, though, we have no clue at this point which versions of most athleticism genes even to choose. There are the rare genes—like EPOR, or myostatin—that alone can have a significant impact on athleticism, but single genes with large impacts have proven the exception. For the foreseeable future, we cannot engineer a genetically ideal athletic specimen. A genetically perfect athlete would simply have to luck into the "right" versions of the genes for her sport.

What are the chances?

Alun Williams, a geneticist at Manchester Metropolitan University in England, was kept awake by that question. So he and his colleague Jonathan P. Folland combed through scientific literature for the twenty-three gene variants that have (so far) been most strongly linked to endurance talent, and then they compiled information about how frequently those gene variants occur in humans.

Some of the variants are found in more than 80 percent of people and others in fewer than 5 percent. Using the gene frequencies, Folland and Williams made statistical projections of how many "perfect" endurance athletes (people with two "correct" versions of the twenty-three genes) walk the planet.

Williams assumed that perfection—even based on the limited number of identified genes—would be uncommon. A Greg LeMond or Chrissie Wellington, after all, is a rare find. But Williams was dumbfounded when he ran the statistical algorithm on his computer and saw that the odds of any single human possessing the perfect set of gene variants was less than one in a quadrillion. To put that in perspective: If you bought twenty lottery tickets per week, you'd have a better chance of winning the Mega Millions *twice in a row* than of hitting that genetic jackpot. Just based on the small number of genes that Folland and Williams included, there is no genetically perfect athlete on earth. Not even close. Given the paltry seven billion people on our planet, chances are that nobody has the ideal endurance profile for more than sixteen of the twenty-three genes. Conversely, an individual is also unlikely to have very few of those endurance genes. Essentially everybody falls in or near the muddled middle, differing by only a handful of genes. It's as if we've all played genetic roulette over and over, moving our chips around, winning sometimes and losing other times, all of us gravitating toward mediocrity. "We're all relatively similar because we're all relying on chance," Williams says.

There are, however, certain elite athletes who do not rely on chance: Thoroughbreds. Because athletic ability involves a complex mix of genes, champion racehorses tend to result from multiple generations of mating among athletic horses. The more genes that are involved in an athletic trait, the more generations of athlete-to-athlete breeding it will likely take to get an offspring that has collected enough of the right gene variants to make the winner's circle. The lone safe bet at the racetrack is that every top horse has racehorses not only for parents but also for grandparents and great-grandparents.

Racehorse breeders have done an outstanding job; the best Thoroughbreds run a mile in a minute and a half. Nonetheless, in many of the world's marquee horse races, the speed of the winners plateaued decades ago. Thoroughbreds may have either reached their physiological terminal velocity or simply run out of new athleticism genes within the breeding population. (Thoroughbreds are relatively in-bred, with more than half of the genes of modern racehorses tracing back to only four individual horses—the Godolphin Arabian, the Darley Arabian, the Byerley Turk, and the Curwen Bay Barb—that traveled from North Africa and the Middle East to England in the late seventeenth and early eighteenth centuries.)

As Pitsiladis put it, to be a world-beater, "you absolutely must choose your parents correctly." He was being facetious, of course, because we can't choose our parents. Nor do humans tend to couple with conscious knowledge of one another's gene variants. We pair up more in the man-ner of a roulette ball that bounces off a few pockets before settling into one of many suitable spots. Williams suggests, hypothetically, that if humanity is to produce an athlete with more "correct" sports genes, one approach is to weight the genetic roulette ball with more lineages in which parents and grandparents are outstanding athletes and thus probably harbor a large number of good athleticism genes. Yao Ming—at 7'5", once the tallest active player in the NBA—was born from Chi-na's tallest couple, a pair of ex–basketball players brought together by the Chinese basketball federation. As Brook Larmer writes in *Operation Yao Ming*: "Two generations of Yao Ming's forebears had been singled out by authorities for their hulking physiques, and his mother and fa-ther were both drafted into the sports system against their will." Still, the witting merger of athletes in pursuit of superstar progeny is rare.

Even that would not guarantee athletic success for any individual offspring of great athletes. In fact, the better the parents are, the less likely it is that the child will be equally good. In any trait that is influ-enced by many genes, it is simply statistically unlikely that a child is going to get as lucky as a very lucky parent. The phrase "regression to the mean" sprang in part from the study of height. Of course, the child of two seven-footers is very likely to be taller than average, but not likely to be as significant of an outlier as his parents. Similarly, the child of two extraordinarily gifted athletes will likely have more of the gene combinations that contribute to athleticism than a randomly selected person, but will be hard-pressed to get as lucky as her mother and father.

In large part, humanity will continue to rely on chance and sports will continue to provide a splendid stage for the fantastic menagerie that is human biological diversity. Amid the pageantry of the Opening Ceremony at the 2016 Olympics in Rio de Janeiro, make sure to look for the extremes of the human physique. The 4'9" gymnast beside the 310-pound shot putter who is looking up at the 6'10" basketball player whose arms are seven and a half feet from fingertip to fingertip. Or the 6'4" swimmer who strides into the Olympic stadium beside his countryman, the 5'9" miler, both men wearing the same length pants.

Our ethnic, geographic, and individual family histories have shaped the genetic information we carry at the nucleus of our every cell and, in turn, our bodies. It is breathtaking to think that, in the truest genetic sense, we are all a large family, and that the paths of our ancestors have left us so wonderfully distinct. In the very last line of his paradigm-shattering *On the Origin of Species*, Charles Darwin says this of his revelation that all the biological variation he sees springs from common ancestry: ". . . from so simple a beginning endless forms most beautiful and most wonderful have been, and are being, evolved."

Because we are each unique, genetic science will continue to show that just as there is no one-size-fits-all medicine, there is no one-size-fits-all training program. If one sport or training method isn't working, it may not be the training. It may be *you*, in the very deepest sense.

Don't be afraid to try something different. Donald Thomas and Chrissie Wellington weren't, and Usain Bolt, after all, had his heart set on cricket stardom.

In the early twentieth century, before the Big Bang of body types, physical education instructors thought the "average" body type was the perfect form for all athletic endeavors. How wrong they were! And now geneticists and physiologists are bolstering evidence that successful practice plans might be as varied as the individuals who would undertake them.

At the end of 2007, the prestigious journal *Science* put "human genetic variation" on its cover as the top scientific breakthrough of the year. As DNA sequencing became cheaper and faster, "researchers are finding out how truly different we are from one another," read the cover story.

To pursue athletic improvement is to embark on a quest in search of the practice plan that suits your inimitable biology. As the HERITAGE Family Study showed, a single exercise program will produce a vast and individualized range of improvement for any particular physical trait. Wonderfully, though, even in HERITAGE there were no "nonresponders" to everything. Sure, there were subjects who saw no improvement in aerobic fitness, but perhaps their blood pressure dropped, or their cholesterol levels improved. Everyone benefits from exercise or sports practice in some unique way. To take part is a journey of self-discovery that, largely, is beyond even the illuminating reach of cutting-edge science.

As J. M. Tanner, the renowned growth expert and world-class hurdler, so elegantly put it: "Everyone has a different genotype. Therefore, for optimal development . . . everyone should have a different environment."

Happy training.

ACKNOWLEDGMENTS

The list of those who deserve thanks is too long for this space. Fortunately, many of their names can be found throughout this book. These are the athletes, scientists, and others who shared their thoughts.

Some, like Yannis Pitsiladis, made time for dozens of interviews. When I followed him to Jamaica, Pitsiladis made sure I could be right there in the operating room as he biopsied a former Jamaican Olympian. I am a richer person for the time I've spent with him.

Physiologists Stephen Roth and Tim Lightfoot scrutinized the entirety of the exercise physiology descriptions in search of errors or imprecisions. Boiling down scientific descriptions while maintaining accuracy is no mean feat, and insofar as I was able to do it, I owe thanks to the patience of dozens of scientists. I also thank my fact-checker, Rebecca Sun, a budding screenwriting talent. If mistakes remain, they are my fault alone.

Every so often I came across a book that humbled me with its depth of research and originality of thought. In two such cases, the authors—J. M. Tanner and Patrick D. Cooper—had passed away. To my regret, I will never have the chance to interview them, but their hard work and liberated thinking will remain in my mind as sources of motivation and courage.

Several colleagues at *Sports Illustrated* deserve special thanks. Without Richard Demak, I doubt I would be writing about sports science for a living. Without Chris Hunt and Craig Neff, I doubt I would've had the space in *SI* for the story that became the seed of this book. Without Terry McDonell and Chris Stone, I doubt I would've had the freedom to work on this book. Without the unfailing encouragement of L. Jon Wertheim and my agent, Scott Waxman, I certainly would have stopped this book before it got started. Thank you, Scott, for foiling my attempt to back out. (Thanks to Farley Chase for work with foreign rights.)

If not for my friendship with Kevin Richards, I most likely would never have turned to sports science writing. Kevin was born in Jamaica and died in Evanston, on a track meet Saturday more than thirteen years ago. I expect that the wound will always be fresh for those of us who ran beside him. I thank Kevin's parents, Gwendolyn and Rupert, and coach David Phillips for their strength. And Kevin Coyne, for teaching me how to write about death and a friend.

In Kenya, I could not have gained access to the places and people (and languages) that I did without Ibrahim Kinuthia, Godfrey Kiprotich, James Mwangi, and Tom and Christopher Ratcliffe. Without Ibrahim and Harun Ngatia, I might still be stuck on the side of the road between Nyahururu and Nairobi looking for a tire that freed itself and bounced over a sheep and away into the brush. (Thanks to the Kenyan children who were kind enough to pluck lug nuts from the dry grass.)

In Jamaica, I thank the University of Technology staff, and particularly Anthony Davis, director of sport, and Colin Gyles, dean of the faculty of science and sport.

In Japan, I thank Noriyuki Fuku and Eri Mikami, of the Tokyo Metropolitan Institute of Gerontology.

In Finland, I thank the Mäntyranta family, but particularly Iiris. And thanks to Elizabeth Newman for helping with phone conversations in Finnish, just as I was beginning to despair in my attempt to track down Eero Mäntyranta.

Puss och kram to my Swedish "family." Especially Kajsa Heinemann, for her friendship in my journeys to Sweden, and also for translating Swedish articles so that I could prepare for my time with Stefan Holm.

On that note, for translation of conversations, papers, or videos, I thank Shiho Takai (Japanese), Alex Von Thun (German), and Veronika Belenkaya (Russian).

My name may be the one on the cover, but if the curtain were removed, many wizards would be visible. Thank you to the staff of the Current imprint at Penguin, particularly marketing director Will Weisser, director of publicity Allison McLean, publicist Jacquelynn Burke, Brittany Wienke and Katie Coe. I reserve special gratitude for editors Adrian Zackheim and Emily Angell. The best way to measure their belief in this project and patience with me is in words: 40,000 of them. That's how many too long I was in the first draft. I also thank Matthew Phillips and Louise Court of Yellow Jersey Press.

Psychologist Drew Bailey's contributions can hardly be overstated. They include tolerating discursive discussions at any time of day, helping with data analysis of NBA bodies, and acting as a personal alert system for new findings that might influence my writing. Genetic science is a moving target, and I could not have tracked it alone. (Thank you, Will Boylan-Pett, for help accessing journals.)

As far as I can tell, my father, Mark Epstein, had scant interest in genetics until I did. Now he is constantly on the lookout for genetics articles and has even had bits of his own genome tested. What greater example can a father provide? My sister, Charna, and brother, Daniel, probably heard "I don't think I can do this" more times than I care to recall. They never believed me. My mother, Eve Epstein, seems always to have known I would write a book. In addition to her help with Swedish translation, her encouragement sustained me. In the course of working on this book, I came across a letter sent from a music teacher to my mother's parents—both of whom fled Germany—when my mother was seven. It reads:

Acknowledgments

I wish to report to you that your daughter is doing exceptional work for the amount of time I have been able to give her. She has an unusual high musical IQ and deserves an expert to give her special attention. I do not have more than a scattered few minutes to show her any special attention and this worries me. In this past twenty years of meeting and working with children, I have never encountered a more alert, exceptional child than Eve. Possibly we could talk it over soon.

Sincerely yours,
Howard Baker

It is a reminder that the requisite nature and nurture are nothing without one another.

Lastly, thank you, Elizabeth. I like to joke to myself that the high pain tolerance of MC1R gene mutants must explain her threshold for my antics. If I ever write another book, I'm sure that one will be dedicated to her too.

NOTES AND
SELECTED CITATIONS

The reporting of this book included hundreds of interviews. In many instances, the interviewees are quoted directly, making the source of that information obvious. In a few cases, high-performance scientists shared with me their data from elite athletes, but asked not to be named, citing the fact that the work is conducted for the purposes of gaining a competitive advantage for a particular team or athlete. Because I do not name scientists or athletes in such cases, I used their data strictly as supporting information for other work.

Additionally, I gained invaluable background at conferences, like the 2010 British Association of Sport and Exercise Sciences conference, and several editions of the American College of Sports Medicine annual meeting. By 2012, I'd been pesky enough in the sports medicine world that I was invited as an ACSM speaker. At that meeting, I also had the profound pleasure of co-organizing, with Yannis Pitsiladis—he of the globe-trotting DNA gathering—an ACSM panel on the nature/nurture of sports expertise. The panel included: Claude Bouchard (the most influential exercise geneticist in the world); K. Anders Ericsson (the man known for the 10,000 hours and the study of deliberate practice); and Philip L. Ackerman (the motor skill acquisition expert who designed the air traffic controller test). Needless to say, the debate was intense, but the dinner afterward amiable and delightful. To me, it was science at its best, contentious and collaborative all at once.

Here I present copious but not comprehensive citations. Throughout the text, it is often easy to track the books and studies I used, as I frequently name the researchers and/or publications. For example, dozens of studies by Janet Starkes and Bruce Abernethy were useful for the first chapter. However, I will not recount their stacks of papers here. These notes are meant to highlight the sources of facts when the source isn't spelled out in the text, and as a detailed point of entry for anyone interested in exploring primary sources. The vast majority of spoken quotes in this book came directly from my interviews. Whenever that is not the case, the source is identified either in the text itself or here.

1

Beat by an Underhand Girl

The Gene-Free Model of Expertise

2   Jennie Finch told me in an interview that she was nervous about Pujols hitting a line drive back at her, and that Bonds refused to allow certain pitches to be filmed. Many of Finch's strikeouts of major leaguers, and Pujols's "I don't want to experience that again" quote, can be found in the DVD titled *MLB Superstars Show You Their Game* (Major League Baseball Productions, 2005).

4   On the problem a human confronts in trying to hit a fastball: Adair, Robert K. *The Physics of Baseball* (3rd ed.). Harper Perennial, 2002. Land, Michael F., and Peter McLeod (2000). "From Eye Movements to Actions: How Batsmen Hit the Ball." *Nature Neuroscience*, 3(12):1340–45. McLeod, P. (1987). "Visual Reaction Time and High-Speed Ball Games." *Perception*, 16(1):49–59.

5   Joe Baker (York University) and Jörg Schorer (University of Muenster) taught me about reaction speed and gave me an occlusion test in which I had to tend a virtual goal against female professional handball players. My results can be inferred from the original chapter 1 title in a first draft of this book: *Beat by a Digital Girl*.

5   For anyone who has ever been told, "Keep your eye on the ball": Bahill, Terry A., and Tom LaRitz (1984). "Why Can't Batters Keep Their Eyes on the Ball?" *American Scientist*, May–June.

6   A sampling of Janet Starkes's work on perceptual expertise and simple reaction time:
Starkes, J. L., and J. Deakin (1984). "Perception in Sport: A Cognitive Approach to Skilled Performance." In W. F. Straub and J. M. Williams, eds. *Cognitive Sports Psychology*, 115–28. Sport Science Intl.
Starkes, J. L. (1987). "Skill in Field Hockey: The Nature of the Cognitive Advantage." *Journal of Sport Psychology*, 9:146–60.

8   De Groot's experiments that laid the foundation for the study of chess expertise:
de Groot, A. D. *Thought and Choice in Chess*. Amsterdam University Press, 2008.

10   Chase and Simon's chunking theory of chess expertise:
Chase, William G., and Herbert A. Simon (1973). "Perception in Chess." *Cognitive Psychology*, (4):55–81.

11   Some of the innovative occlusion work by Bruce Abernethy and colleagues:
Abernethy, B., et al. (2008). "Expertise and Attunement to Kinematic Constraints." *Perception*, 37(6):931–48.
Mann, David L., et al. (2010). "An Event-Related Visual Occlusion Method for Examining Anticipatory Skill in Natural Interceptive Tasks." *Behavior Research Methods*, 42(2):556–62.

Muller, S., et al. (2006). "How do World-Class Cricket Batsmen Anticipate a Bowler's Intention?" *Quarterly Journal of Experimental Psychology*, 59(10):2162–86.

12   The visual reaction speed of Muhammad Ali, and how Ali's test results were initially misportrayed:
Kamin, Leon J., and Sharon Grant-Henry (1987). "Reaction Time, Race, and Racism." *Intelligence*, 11:299–304.

12   The perceptual expertise of basketball rebounding:
Aglioti, Salvatore M., et al. (2008). "Action Anticipation and Motor Resonance in Elite Basketball Players." *Nature Neuroscience*, 11(9):1109–16.

13   Psychologist Richard Abrams provided several of the results of Washington University's 2006 testing of Pujols:
http://news.wustl.edu/news/pages/7535.aspx.

13   Detailed background on the study of skill expertise in sports:
Starkes, Janet L., and K. Anders Ericsson, eds. *Expert Performance in Sports: Advances in Research in Sport Expertise*. Human Kinetics, 2003.

13   Practice at a specific task changes the brain and leads to automation:
Duerden, Emma G., and Danièle Laverdure-Dupont (2008). "Practice Makes Cortex." *The Journal of Neuroscience*, 28(35):8655–57.
Squire, Larry, and Eric Kandel. *Memory: From Mind to Molecules* (chap. 9). Macmillan, 2000.
Van Raalten, Tamar R., et al. (2008). "Practice Induces Function-Specific Changes in Brain Activity." *PLoS ONE*, 3(10):e3270.

13   Familiarity with a familiar mode of exercise influences brain activity. A study of interest:
Brümmer, V., et al. (2001). "Brain Cortical Activity Is Influenced by Exercise Mode and Intensity." *Medicine & Science in Sports & Exercise*, 43(10):1863–72.

14   The best primer on the modern study of expertise, from chess to surgery to writing, with emphasis on "software":
Ericsson, K. Anders, et al., eds. *The Cambridge Handbook of Expertise and Expert Performance*. Cambridge University Press, 2006.

2

A Tale of Two High Jumpers

(Or: 10,000 Hours Plus or Minus 10,000 Hours)

18   Dan McLaughlin's progress can be followed at: thedanplan.com.

21   Numerous chess studies by Campitelli and/or Gobet were used in reporting, but these were the most central:
Campitelli, Guillermo, and Fernand Gobet (2008). "The Role of Practice in Chess: A Longitudinal Study." *Learning and Individual Differences*, 18(4):446–58.

Gobet, F., and G. Campitelli (2007). "The Role of Domain-Specific Practice, Handedness, and Starting Age in Chess." *Developmental Psychology,* 43(1):159–72.

Gobet, Fernand, and Herbert A. Simon (2000). "Five Seconds or Sixty? Presentation Time in Expert Memory." *Cognitive Science,* 24(4):651–82.

22 The paper in which K. Anders Ericsson writes that Gladwell "misconstrued" his conclusion:

Ericsson, K. Anders (2012). "Training History, Deliberate Practise and Elite Sports Performance: An Analysis in Response to Tucker and Collins Review—What Makes Champions?" *British Journal of Sports Medicine,* Oct. 30 (ePub ahead of print).

23 Holm's personal Web site (scholm.com) is a testament to a lifelong obsession with high jump (and Legos).

29 Photos of Thomas's first competition (in baggy shorts) are preserved here: http://www.polevaultpower.com/forum/viewtopic.php?f=32&t=7161&sid=e68562cf62585697482f1ec91c086165.

29 Most details come from Thomas himself and competition records, but the quote by Thomas's cousin that Thomas "doesn't know that a track goes around in a circle," and Clayton's "didn't know how to warm up" quote both originally appeared in a 2007 press release issued by the U.S. Track and Field and Cross Country Coaches Association titled: "An Improbable Leap into the Limelight."

31 YouTube has video of Thomas's world championship win: http://www.youtube.com/watch?v=yzmPtZyuo4s.

32 Johnny Holm's "buffoon" quote appeared in the Swedish publication *Sport Expressen* on August 30, 2007. It can be found here: http://www.expressen.se/sport/friidrott/han-ar-en-javla-pajas/.

32 The NHK documentary on Holm and Thomas—the title roughly translates to "Inside the Top Athlete's Body"—is brilliant.

33 A good example of the tremendous range of practice hours accumulated by competitors of similar ability:

Baker, Joseph, Jean Côté, and Janice Deakin (2005). "Expertise in Ultra-Endurance Triathletes: Early Sport Improvement, Training Structure, and the Theory of Deliberate Practice." *Journal of Applied Sport Psychology,* 17:64–78.

34 Among papers that chronicle the number of practice hours that elite athletes accumulate:

Baker, Joseph, Jean Côté, and Bruce Abernethy (2003). "Sport-Specific Practice and the Development of Expert Decision-Making in Team Ball Sports." *Journal of Applied Sport Psychology,* 15:12–25.

Helsen, W. F., J. L. Starkes, and N. J. Hodges (1998). "Team Sports and the Theory of Deliberate Practice." *Journal of Sport & Exercise Psychology,* 20:12–34.

Hodges, N. J., and J. L. Starkes (1996). "Wrestling with the Nature of Expertise: A Sport Specific Test of Ericsson, Krampe and Tesch-Römer's (1993) theory of 'deliberate practice.'" *International Journal of Sport Psychology*, 27:400–24.

Williams, Mark A., and Nicola J. Hodges, eds. *Skill Acquisition in Sport: Research, Theory and Practice* (chap. 11). Routledge, 2004.

34 On the 28 percent of Australian athletes who reached the international level after only four years:

Bullock, Nicola, et al. (2009). "Talent Identification and Deliberate Programming in Skeleton: Ice Novice to Winter Olympian in 14 Months." *Journal of Sports Sciences*, 27(4):397–404.

Oldenziel, K., F. Gagne, and J. P. Gulbin (2004). "Factors Affecting the Rate of Athlete Development from Novice to Senior Elite: How Applicable Is the 10-Year Rule?" Pre-Olympic Congress, Athens. (Summary here: http://cev .org.br/biblioteca/factors-affecting-the-rate-of-athlete-development-from -novice-to-senior-elite-how-applicable-is-the-10-year-rule/.)

35 Thorndike, Edward L. (1908). "The Effect of Practice in the Case of a Purely Intellectual Function." *American Journal of Psychology*, 19:374–384.

37 Even in darts, accumulated practice explains a small portion of variance in performance after fifteen years:

Duffy, Linda J., Bahman Baluch, and K. Anders Ericsson (2004). "Dart Performance as a Function of Facets of Practice Amongst Professional and Amateur Men and Women Players." *International Journal of Sport Psychology*, 35:232–45.

3

Major League Vision and the Greatest Child Athlete Sample Ever

The Hardware *and* Software Paradigm

38 Rosenbaum recounts some of his Dodgers work in his book *Beware of GUS: Government-University Symbiosis*. Lulu.com, 2010.

39 The main paper with data from the Dodgers (Daniel M. Laby kindly provided additional data):

Laby, Daniel M., et al. (1996). "The Visual Function of Professional Baseball Players." *American Journal of Ophthalmology*, 122:476–85.

39 The theoretical limit of human visual acuity:

Applegate, Raymond A. (2000). "Limits to Vision: Can We Do Better Than Nature?" *Journal of Refractive Surgery*, 16: S547–51.

39   On the range of human cone density:
     Curcio, Christine A., et al. (1990). "Human Photoreceptor Topography."
     Journal of Comparative Neurology, 292:497–523.
40   Piazza picked as a favor to his father:
     Whiteside, Kelly. "A Piazza with Everything." *Sports Illustrated*, July 5, 1993.
40   The China and India vision studies:
     Nangia, Vinay, et al. (2011). "Visual Acuity and Associated Factors:
     The Central India Eye and Medical Study." *PLoS ONE*, 6(7):e22756.
     Xu, L., et al. (2005). "Visual Acuity in Northern China in an Urban and
     Rural Population: The Beijing Eye Study." *British Journal of Ophthalmology*,
     89:1089–93.
40   Studies of visual acuity in young people, including Swedish teenagers:
     Frisén, L., and M. Frisén (1981). "How Good Is Normal Visual Acuity? A
     Study of Letter Acuity Thresholds as a Function of Age." *Albrecht von Graefes
     Archiv für klinische und experimentelle Ophthalmologie*, 215(3):149–57.
     Ohlsson, Josefin, and Gerardo Villarreal (2005). "Normal Visual Acuity in
     17–18 Year Olds." *Acta Ophthalmologica Scandinavia*, 83:487–91.
41   As a group, hitters begin to decline at age twenty-nine:
     Fair, Ray C. (2008). "Estimated Age Effects in Baseball." *Journal of Quantitative Analysis in Sports*, 4(1):1.
41   Ted Williams on his own vision:
     Williams, Ted, and John W. Underwood. *My Turn at Bat: The Story of My Life.*
     Simon and Schuster, 1988, p. 93–94.
42   Keith Hernandez's quote is from his commentary on SNY during the sixth
     inning of the Mets game against the Nationals on April 10, 2012.
42   Virtual-reality batting studies:
     Gray, Rob (2002). "Behavior of College Baseball Players in a Virtual Batting Task." *Journal of Experimental Psychology: Human Perception and Performance*, 28(5):1131–48.
     Hyllegard, R. (1991). "The Role of Baseball Seam Pattern in Pitch Recognition." *Journal of Sport & Exercise Psychology*, 13:80–84.
42   Most tennis pros have outstanding visual acuity, but a few have average vision:
     Fremion, Amy S., et al. (1986). "Binocular and Monocular Visual Function
     in World Class Tennis Players." *Binocular Vision*, 1(3):147–54.
43   Muhammad Ali's reaction speed:
     Kamin, Leon J., and Sharon Grant-Henry (1987). "Reaction Time, Race,
     and Racism." *Intelligence*, 11:299–304.
43   Visual-acuity of Olympians:
     Laby, Daniel M., David G. Kirschen, and Paige Pantall (2011). "The Visual
     Function of Olympic-Level Athletes—An Initial Report." *Eye & Contact
     Lens*, Mar. 3 (ePub ahead of print).
43   Depth perception and catching skills:

Mazyn, Liesbeth I. N., et al. (2004). "The Contribution of Stereo Vision to One-Handed Catching." *Experimental Brain Research*, 157:383–90.

Mazyn, Liesbeth I. N., et al. (2007). "Stereo Vision Enhances the Learning of a Catching Skill." *Experimental Brain Research*, 179:723–26.

44　Emory study of youth baseball/softball players:

Boden, Lauren M., et al. (2009). "A Comparison of Static Near Stereo Acuity in Youth Baseball/Softball Players and Non–Ball Players." *Optometry*, 80:121–25.

45　Schneider's tennis study is published only in German:

Schneider, W., K. Bös, and H. Rieder (1993). "Leistungsprognose bei jugendlichen Spitzensportlern [Performance prediction in adolescent top tennis players]." In: J. Beckmann, H. Strang, and E. Hahn, eds., *Aufmerksamkeit und Energetisierung*. Göttingen: Hogrefe.

46　Graf's training with Germany's Olympic track team is mentioned in her husband's memoir:

Agassi, Andre. *Open*. Vintage, 2010 (Kindle e-book).

46　An introduction to the Groningen talent studies:

Elferink-Gemser, Marije T., et al. (2004). "The Marvels of Elite Sports: How to Get There?" *British Journal of Sports Medicine*, 45:683–84.

Elferink-Gemser, Marije T., and Chris Visscher. "Chapter 8: Who Are the Superstars of Tomorrow? Talent Development in Dutch Soccer." In: Joseph Baker, Steve Cobley, and Jörg Schorer, eds. *Talent Identification and Development in Sport: International Perspectives*. Routledge, 2011.

50　The difference in practice hours between Belgian and Dutch field hockey players:

van Rossum, Jacques H. A. "Chapter 37: Giftedness and Talent in Sport." In: L. V. Shavinina, ed. *International Handbook on Giftedness*. Springer, 2009.

50　Diverse, rather than specialized sports experience can lead to the attainment of expertise in certain sports:

Baker, Joseph (2003). "Early Specialization in Youth Sport: A Requirement for Adult Expertise?" *High Ability Studies*, 14(1):85–94.

Baker, Joseph, Jean Côté, and Bruce Abernethy (2003). "Sport-Specific Practice and the Development of Expert Decision-Making in Team Ball Sports." *Journal of Applied Sport Psychology*, 15:12–25.

52　Discussion of the "speed plateau":

Schiffer, Jürgen (2011). "Training to Overcome the Speed Plateau." *New Studies in Athletics*, 26(1/2):7–16.

53　Tiger Woods, on his desire to play:

Verdi, Bob. "The Grillroom: Tiger Woods." *Golf Digest*. January 1, 2000, 51(1):132.

53　Tiger could balance on his father's palm at six months:

Smith, Gary. "The Chosen One." *Sports Illustrated*. December 23, 1996.

4

Why Men Have Nipples

56   The best read on the travails of María José Martínez-Patiño was written by
Martínez-Patiño herself:
Martínez-Patiño, María José (2005). "Personal Account: A Woman Tried
and Tested." *Lancet*, 366:S38.

59   *U.S. News & World Report* surveyed Americans on whether female athletes
would soon beat male athletes:
Holden, Constance (2004). "An Everlasting Gender Gap?" *Science*, 305:
639–40.

59   The papers suggesting that women will outrun men:
Beneke, R., R. M. Leithäuser, and M. Doppelmayr (2005). "Women Will Do
It in the Long Run." *British Journal of Sports Medicine*, 39:410.
Tatem, Andrew J., et al. (2004). "Momentous Sprint at the 2156 Olympics?
Women Sprinters Are Closing the Gap on Men and May One Day Overtake
Them." *Nature*, 431:525.
Whipp, Brian J., and Susan A. Ward (1992). "Will Women Soon Outrun
Men?" *Nature*, 355:25.

60   Men out-throw women by three standard deviations, and the gap starts
before sports participation:
Thomas, Jerry R., and Karen E. French. "Gender Differences Across Age in
Motor Performance: A Meta-Analysis." *Psychological Bulletin*, 98(2):260–82.

61   Background on sexual differentiation (particularly chapter 1):
Baron-Cohen, Simon, Svetlana Lutchmaya, and Rebecca Knickmeyer. *Prenatal Testosterone in Mind: Amniotic Fluid Studies*. The MIT Press, 2004.

61   David C. Geary's book *Male, Female: The Evolution of Human Sex Differences*, 2nd ed., American Psychological Association, 2010, is a fascinating
read and the main resource for facts about sex differences in this chapter
(example: boys develop longer forearms than girls while still in the womb;
30 percent of hunter-gatherer men died at the hands of other men; sex differences in upper-body strength). This compilation of one hundred years
of studies of sex differences was also used:
Ellis, Lee, et al. *Sex Differences: Summarizing More Than a Century of Scientific
Research*. Psychology Press, 2008.

61   The male/female throwing gap, and throwing skill in Australian Aboriginal children:
Thomas, Jerry R., et al. (2010). "Developmental Gender Differences for
Overhand Throwing in Australian Aboriginal Children." *Research Quarterly for Exercise and Sport*, 81(4):1–10.

62   Sexual selection and physical competition in humans and other animals,
and targeting skill differences:

Puts, David A. (2010). "Beauty and the Beast: Mechanisms of Sexual Selection in Humans." *Evolution and Human Behavior*, 31:157–75.

62 Targeting skills of females who are exposed to higher than normal levels of testosterone prenatally:
Hines, M., et al. (2003). "Spatial Abilities Following Prenatal Androgen Abnormality: Targeting and Mental Rotations Performance in Individuals with Congenital Adrenal Hyperplasia." *Psychoneuroendocrinology*, 28(8):1010–26.

62 Despite the throwing gap, highly trained women will out-throw untrained men:
Schorer, Jörg, et al. (2007). "Identification of Interindividual and Intraindividual Movement Patterns in Handball Players of Varying Expertise Levels." *Journal of Motor Behavior*, 39(5):409–21.

62 Analysis of the elite performance gap in track and field and swimming:
Thibault, Valérie, et al. (2010). "Women and Men in Sport Performance: The Gender Gap Has Not Evolved Since 1983." *Journal of Sports Science and Medicine*, 9:214–23.

62 Sex differences in ultraendurance races, starting on p. 682 of a book known to a generation of runners:
Noakes, Timothy D. *Lore of Running* (4th ed.). Human Kinetics, 2002.

63 The widening running gap between men and women:
Denny, Mark W. (2008). "Limits to Running Speed in Dogs, Horses and Humans." *The Journal of Experimental Biology*, 211:3836–49.
Holden, Constance (2004). "An Everlasting Gender Gap?" *Science*, 305: 639–40.

65 Sex differences in skeletal growth and proportions:
Malina, Robert, Claude Bouchard, and Oded Bar-Or. *Growth, Maturation & Physical Activity* (2nd ed.). Human Kinetics, 2003.
Malina, Robert M. "Part Five: Post-natal Growth and Maturation." In: Stanley J. Ulijaszek, et al. eds. *The Cambridge Encyclopedia of Human Growth and Development*. Cambridge University Press, 1998.
Morgenthal, Paige A., and Diane N. Resnick. "Chapter 14: The Female Athlete: Current Concepts." In: Robert D. Mootz and Kevin McCarthy, eds., *Sports Chiropractic*. Jones & Bartlett Learning, 1999.

65 A table listing basic physical differences between the sexes that are relevant to athleticism is on p. 176 of:
Abernethy, Bruce, et al. *The Biophysical Foundations of Human Movement* (2nd ed.). Human Kinetics, 2004.

66 Physical competition depends on the area inhabited by the organism:
Puts, David A. (2010). "Beauty and the Beast: Mechanisms of Sexual Selection in Humans." *Evolution and Human Behavior*, 31:157–75.

67 Studies that document the larger number of female than male ancestors of modern humans are numerous, but a summary can be found in Geary's *Male, Female: The Evolution of Human Sex Differences*, on pp. 234–35.

67   The "Genghis Khan paper":
Zerjal, T., et al. (2003). "The Genetic Legacy of the Mongols." *American Journal of Human Genetics*, 72:717–21.

67   Meta-analysis of the pre- and postpuberty gap in athletic skills between males and females ages two to twenty:
Thomas, Jerry R., and Karen E. French. "Gender Differences Across Age in Motor Performance: A Meta-Analysis." *Psychological Bulletin*, 98(2):260–82.

67   Prior to puberty, boys and girls do not differ in height or muscle and bone mass:
Gooren, Louis J. (2008). "Olympic Sports and Transsexuals." *Asian Journal of Andrology*. 10(3):427–32.

68   Age-related changes in boys and girls for a range of physical skills—throwing, sprinting—are in chapter 11 of:
Malina, Robert, Claude Bouchard, and Oded Bar-Or. *Growth, Maturation & Physical Activity* (2nd ed.). Human Kinetics, 2003.

68   Discussion of physical characteristics, including body fat, of female marathoners:
Christensen, Carol L., and R. O. Ruhling (1983). "Physical Characteristics of Novice and Experienced Women Marathon Runners." *British Journal of Sports Medicine*, 17(3):166–71.

68   Discussion of body size and performance in developing gymnasts:
Claessens, Albrecht L. (2006). "Maturity-Associated Variation in the Body Size and Proportions of Elite Female Gymnasts 14–17 Years of Age." *European Journal of Pediatrics*, 165:186–92.
Malina, R. M. (1994). "Physical Growth and Biological Maturation of Young Athletes." *Exercise and Sport Sciences Reviews*, 22:389–433.

69   A captivating look into the East German doping program:
Ungerleider, Steven. *Faust's Gold: Inside the East German Doping Machine*. Thomas Dunne Books, 2001.

70   Two excellent reviews of intersex conditions in Olympians:
Ritchie, Robert, John Reynard, and Tom Lewis (2008). "Intersex and the Olympic Games." *Journal of the Royal Society of Medicine*, 101:395–99.
Tucker, Ross, and Malcolm Collins (2009). "The Science and Management of Sex Verification in Sport." *South African Journal of Sports Medicine*, 21(4):147–150.

70   The male and female ranges of testosterone come from interviews with endocrinologists and lab reference ranges. The testosterone reference range varies slightly by lab. Quest Diagnostics provides a male range of 241–827 nanograms of testosterone per deciliter of blood. The Mayo Clinic provides a similar range: http://www.mayomedicallaboratories .com/test-catalog/Clinical+and+Interpretive/8508.

71   Seven female athletes at the Atlanta Olympics who were found to have an SRY gene:

Wonkam, Ambroise, Karen Fieggen, and Raj Ramesar (2010). "Beyond the Caster Semenya Controversy." *Journal of Genetic Counseling,* 19(6):545–548.

71 The prevalence of a Y chromosome in female competitors over five Olympics: Foddy, Bennett, and Julian Savulescu (2011). "Time to Re-evaluate Gender Segregation in Athletics?" *British Journal of Sports Medicine,* 45(15):1184–88.

71 Rates of complete androgen insensitivity syndrome:
Galani, Angeliki, et al. (2008). "Androgen Insensitivity Syndrome: Clinical Features and Molecular Defects." *Hormones,* 7(3):217–29.

71 Among the studies that document tall stature and masculine skeletal ratios in women with AIS:
Han T. S., et al. (2008). "Comparison of Bone Mineral Density and Body Proportions Between Women with Complete Androgen Insensitivity Syndrome and Women with Gonadal Dysgenesis." *European Journal of Endocrinology,* 159:179–85.
Zachmann, M., et al. (1986). "Pubertal Growth in Patients with Androgen Insensitivity: Indirect Evidence for the Importance of Estrogens in Pubertal Growth of Girls." *Journal of Pediatrics,* 108:694–97.

71 Androgen insensitivity only "scratches the surface" of intersex conditions in sports:
Foddy, Bennett, and Julian Savulescu (2011). "Time to Re-Evaluate Gender Segregation in Athletics?" *British Journal of Sports Medicine,* 45(15):1184–88.

72 The testosterone levels of elite female athletes:
Cook, C. J., et al. (2012). "Comparison of Baseline Free Testosterone and Cortisol Concentrations Between Elite and Non-Elite Athletes." *American Journal of Human Biology,* 24(6):856–58.

72 Female netball players with higher testosterone self-select greater workloads:
Cook, C. J., and C. M. Beaven (2013). "Salivary Testosterone is Related to Self-Selected Training Load in Elite Female Athletes." *Physiology & Behavior,* 116-117C:8-12 (ePub ahead of print).

74 Men's hearts get bigger more rapidly:
Kolata, Gina. "Men, Women and Speed. 2 Words: Got Testosterone?" *New York Times,* August 22, 2008.

5

The Talent of Trainability

78 In addition to interviews with Ryun, his book, *In Quest of Gold: The Jim Ryun Story,* written with Mike Phillips, gives a detailed account of his emergence in track and field and is the source of quotes from Ryun's parents and his own writing.

79   The HERITAGE Family Study has produced more than one hundred jour-
      nal articles. The HERITAGE papers most central to this chapter:
      Bouchard, Claude, et al. (1999). "Familial Aggregation of VO$_2$max Re-
      sponse to Exercise Training: Results from the HERITAGE Family Study."
      Journal of Applied Physiology, 87:1003–8.
      Bouchard, Claude, et al. (2011). "Genomic Predictors of the Maximal O2
      Uptake Response to Standardized Exercise Training Programs." *Journal of
      Applied Physiology*, 10(5):1160–70.
      Rankinen, T., et al. (2010). "CREB1 Is a Strong Genetic Predictor of
      the Variation in Exercise Heart Rate Response to Regular Exercise:
      The HERITAGE Family Study." *Circulation: Cardiovascular Genetics*, 3(3):
      294–99.
      Timmons, James A., et al. (2010). "Using Molecular Classification to Pre-
      dict Gains in Maximal Aerobic Capacity Following Endurance Exercise
      Training in Humans." *Journal of Applied Physiology*, 108:1487–96.

79   A layman's introduction to the HERITAGE Family Study can be found here:
      Roth, Stephen M. *Genetics Primer for Exercise Science and Health*. Human Ki-
      netics, 2007.

83   The independent scientific commentary on the twenty-nine-gene expres-
      sion signature:
      Bamman, Marcas M. (2010). "Does Your (Genetic) Alphabet Soup Spell
      'Runner'?" *Journal of Applied Physiology*, 108:1452–53.

84   Data from Miami's GEAR study were kindly shared by members of the re-
      search team, particularly: Pascal J. Goldschmidt (dean, Miller School of
      Medicine, University of Miami); Margaret A. Pericak-Vance (director, Mi-
      ami Institute of Human Genomics); Jeffrey Farmer (GEAR project man-
      ager); Evadnie Rampersaud (director, Division of Genetic Epidemiology in
      the Center for Genetic Epidemiology and Statistical Genetics, John P.
      Hussman Institute for Human Genomics).

91   The "naturally fit six" study:
      Martino, Marco, Norman Gledhill, and Veronica Jamnik (2002). "High
      VO$_2$max with No History of Training Is Primarily Due to High Blood Vol-
      ume." *Medicine & Science in Sports & Exercise*, 34(6):966–71.

94   Wellington's "near impossible task":
      "Wellington Wins World Ironman Championships." Britishtriathlon.org,
      October 14, 2007.

96   Andrew Wheating's entry into track and field is described here:
      Layden, Tim. "Off to a Blazing Start." *Sports Illustrated*, September 20, 2010.

96   Alberto Juantorena recounts his switch from basketball to track here:
      Sandrock, Michael. *Running with the Legends*. Human Kinetics, 1996, p. 204.

97   Jack Daniels's five-year study of Jim Ryun:
      Daniels, Jack (1974). "Running with Jim Ryun: A Five-Year Study." *The Physi-
      cian and Sportsmedicine*, 2:63–67.

98   Study of Japanese junior athletes:
     Murase, Yutaka, et al. (1981). "Longitudinal Study of Aerobic Power in Superior Junior Athletes." *Medicine & Science in Sports & Exercise,* 13(3):180–84.

6

Superbaby, Bully Whippets, and the Trainability of Muscle

100   The original Superbaby paper:
      Schuelke, Marcus, et al. (2004). "Myostatin Mutation Associated with Gross Muscle Hypertrophy in a Child." *New England Journal of Medicine,* 350:2682–88.
101   The first description of myostatin in scientific literature:
      McPherron, Alexandra C., Ann M. Lawler, and Se-Jin Lee (1997). "Regulation of Skeletal Muscle Mass in Mice by a New TGF-β Superfamily Member." *Nature,* 387(6628):83–90.
102   The myostatin mutation found in cattle:
      McPherron, Alexandra C., and Se-Jin Lee (1997). "Double Muscling in Cattle Due to Mutations in the Myostatin Gene." *Proceedings of the National Academy of Sciences,* 94:12457–61.
103   Whippets and the myostatin mutation:
      Mosher, Dana S., et al. (2007). "A Mutation in the Myostatin Gene Increases Muscle Mass and Enhances Racing Performance in Heterozygote Dogs." *PLoS ONE,* 3(5):e79.
104   Myostatin gene predicts sprinting ability and earnings in horses:
      Hill, Emmeline W., et al. (2010). "A Sequence Polymorphism in MSTN Predicts Sprinting Ability and Racing Stamina in Thoroughbred Horses." *PLoS ONE,* 5(1):e8645.
104   The impact of variations in the myostatin gene on athletic performance in animals:
      Lee, Se-Jin (2007). "Sprinting Without Myostatin: A Genetic Determinant of Athletic Prowess." *Trends in Genetics,* 23(10):475–77.
      Lee, Se-Jin (2010). "Speed and Endurance: You Can Have It All." *Journal of Applied Physiology,* 109:621–22.
105   Myostatin-inhibiting molecule increased mouse muscle 60 percent in two weeks:
      Lee, Se-Jin, et al. (2005). "Regulation of Muscle Growth by Multiple Ligands Signaling Through Activin Type II Receptors." *Proceedings of the National Academy of Sciences,* 102(50):18117–22.
105   Pharmaceutical companies are testing drugs that inhibit myostatin in humans:

Attie, Kenneth M., et al. (2012). "A Single Ascending-Dose Study of Muscle Regulator ACE-031 in Health Volunteers." *Muscle & Nerve*, August 1 (ePub ahead of print).

106   H. Lee Sweeney on his IGF-1 work and the future prospect of gene doping: Sweeney, H. Lee (2004). "Gene Doping." *Scientific American*, (July 2004): 63–69.

107   Studies by University of Alabama–Birmingham's Core Muscle Research Laboratory and the Veterans Affairs Medical Center: Bamman, Marcas M., et al. (2007). "Cluster Analysis Tests the Importance of Myogenic Gene Expression During Myofiber Hypertrophy in Humans." *Journal of Applied Physiology*, 102:2232–39.
Petrella, John K., et al. (2008). "Potent Myofiber Hypertrophy During Resistance Training in Humans Is Associated with Satellite Cell-Mediated Myonuclear Addition: A Cluster Analysis." *Journal of Applied Physiology*, 104: 1736–42.

108   GEAR study data was generously shared by members of the University of Miami research team.

108   After twelve weeks, strength gains ranged from 0 percent to 250 percent: Hubal, M. J., et al. (2005). "Variability in Muscle Size and Strength Gain After Unilateral Resistance Training." *Medicine & Science in Sports & Exercise*, 37(6):964–72.

109   Muscle contraction speed limits human sprinting: Weyand, Peter G., et al. (2010). "The Biological Limits to Running Speed Are Imposed from the Ground Up." *Journal of Applied Physiology*, 108(4):950–61.

109   An accessible introduction to muscle fiber types, with a chart showing typical proportions: Andersen, Jesper L., et al. (2007). "Muscle, Genes and Athletic Performance." In: Editors of *Scientific American*, ed. *Building the Elite Athlete*. Scientific American.

110   Two of the most famous studies of muscle fiber proportions in athletes: Costill, D. L., et al. (1976). "Skeletal Muscle Enzymes and Fiber Composition in Male and Female Track Athletes." *Journal of Applied Physiology*, 40(2):149–54.
Fink, W. J., D. L. Costill. and M. L. Pollock (1977). "Submaximal and Maximal Working Capacity of Elite Distance Runners. Part II: Muscle Fiber Composition and Enzyme Activities." *Annals of the New York Academy of Sciences*, 301:323–27.

110   An excellent and freely available primer on muscle fiber types: Zierath, Juleen R., and John A. Hawley. "Skeletal Muscle Fiber Type: Influence on Contractile and Metabolic Properties." *PLoS Biology*, 2(10):e348.

110   Frank Shorter's biopsied calf muscle can be viewed for free online in fig. 2 of this paper: Zierath, Juleen R., and John A. Hawley. "Skeletal Muscle Fiber Type: Influence on Contractile and Metabolic Properties." *PLoS Biology*, 2(10):e348.

110   Eight hours a day of electrical stimulation did not change slow-twitch fiber proportions:
Simoneau, Jean-Aimé, and Claude Bouchard (1995). "Genetic Determinism of Fiber Type Proportion in Human Skeletal Muscle." *The FASEB Journal*, 9:1091–95.

110   The review, coauthored by Jesper Anderson, addressing the impact of training on muscle fibers:
Andersen, J. L., and P. Aagaard (2010). "Effects of Strength Training on Muscle Fiber Types and Size: Consequences for Athletes Training for High-Intensity Sport." *Scandinavian Journal of Medicine & Science in Sports*, 20(Suppl. 2):32–38.

110   The Russian study correlating endurance genes and muscle fiber proportions:
Ahmetov, Ildus I. (2009). "The Combined Impact of Metabolic Gene Polymorphisms on Elite Endurance Athlete Status and Related Phenotypes." *Human Genetics*, 126(6):751–61.

7

The Big Bang of Body Types

114   Winner-take-all markets with discussion of the impact of technology:
Frank, Robert H. *Luxury Fever: Money and Happiness in an Era of Excess.* Free Press, 1999 (Kindle e-book).

115   The joint speed of Jesse Owens was similar to that of Carl Lewis:
Schechter, Bruce. "How Much Higher? How Much Faster?" In: Editors of *Scientific American*, eds. Building the Elite Athlete. Scientific American, 2007.

115   The quote regarding the perfect form of man appears here:
Sargent, D. A. (1887). "The Physical Characteristics of the Athlete." *Scribner's Magazine*, 2(5):558.

116   Norton and Olds have written extensively on the changing bodies in the elite athlete pool. Here are two of the best compilation papers, from which many of the sport-specific examples in this chapter were drawn:
Norton, Kevin, and Tim Olds (2001). "Morphological Evolution of Athletes Over the 20th Century: Causes and Consequences." *Sports Medicine*, 31(11):763–83.
Olds, Timothy. "Chapter 9: Body Composition and Sports Performance." In: Ronald J. Maughan, ed. *The Olympic Textbook of Science in Sport*, Blackwell Publishing, 2009.

117   Very tall women are 191 times more likely to make an Olympic final than very small women:
Khosla, T., and V. C. McBroom (1988). "Age, Height and Weight of Female Olympic Finalists." *British Journal of Sports Medicine*, 19:96–99.

119 Norton and Olds coedited the textbook *Anthropometrica* (UNSW Press, 2004), the definitive introduction to the measurement of body types in sports. Chapter 11, "Anthropometry and Sports Performance," is a treasure trove of information, from the rapid change in the height of high jumpers after the introduction of the Fosbury flop, to graphs showing how the bodies of world record holders vary according to the distance they run.

119 Heat dissipation and body size of runners:
O'Connor, Helen, Tim Olds, and Ronald J. Maughan (2007). "Physique and Performance for Track and Field Events." *Journal of Sports Sciences*, 25(S2):S49–60.

120 The effect of core temperature on effort (and the impact of amphetamines):
Roelands, Bart, et al. (2008). "Acute Norepinephrine Reuptake Inhibition Decreases Performance in Normal and High Ambient Temperature." *Journal of Applied Physiology*, 105:206–12.
Tucker, Ross (2009). "The Anticipatory Regulation of Performance: The Physiological Basis for Pacing Strategies and the Development of a Perception-Based Model for Exercise Performance." *British Journal of Sports Medicine*, 43:392–400.

120 Heat dissipation discussion specifically with respect to Paula Radcliffe:
Schwellnus, Martin P., ed. *The Olympic Textbook of Medicine in Sport*. Wiley, 2008, p. 463.

120 The famous 1968 Mexico City Olympics study of body types:
de Garay, Alfonso L., Louise Levine, and J. E. Lindsay Carter, eds. *Genetic and Anthropological Studies of Olympic Athletes*. Academic Press, 1974.

121 Michael Phelps's short inseam:
McMullen, Paul. "Measure of a Swimmer: From Flipper Feet to a Long Trunk, Phelps Represents a One-Man Body Shop of What a Swimmer Should Be." *Baltimore Sun*, March 9, 2004.

122 Salary gap between average workers and pro athletes (updated using figures from the U.S. Census Bureau):
Olds, Timothy. "Chapter 9: Body Composition and Sports Performance." In: Ronald J. Maughan, ed. *The Olympic Textbook of Science in Sport*. Blackwell Publishing, 2009.

122 The GIANT Consortium study:
Willer, C. J., et al. (2009). "Six New Loci Associated with Body Mass Index Highlight a Neuronal Influence on Body Weight Regulation." *Nature Genetics*, 41(1):25–34.

123 Researchers in the United States and Finland have found that a high proportion of fast-twitch muscle fibers decreases fat burning and increases blood pressure and risk of heart disease:
Hernelahti, Miika, et al. (2008). "Muscle Fiber-Type Distribution as a Predictor of Blood Pressure: A 19-Year Follow-Up Study." *Hypertension*, 45(5):1019–23.

Kujala, Urho M., and Heikki O. Tikkanen (2001). "Disease-Specific Mortality Among Elite Athletes." *JAMA*, 285(1):44.

Tanner, Charles J., et al. (2002). "Muscle Fiber Type Is Associated with Obesity and Weight Loss." *American Journal of Physiology—Endocrinology and Metabolism*, 282:E1191–96.

124 Francis Holway graciously shared spreadsheets of his data on the body measurements of athletes.

124 Cowgill on innate skeletal differences:

Cowgill, L. W. (2010). "The Ontogeny of Holocene and Late Pleistocene Human Postcranial Strength." *American Journal of Physical Anthropolgy*, 141(1):16–37.

126 Tanner's quote comes from:

Tanner, J. M. *Fetus into Man: Physical Growth from Conception to Maturity* (revised and enlarged edition). Harvard University Press, 1990.

8

The Vitruvian NBA Player

129 Dennis Rodman confirmed his rapid height growth in an interview, but his book is the most colorful account and provided his quotes:

Rodman, Dennis. *Bad as I Wanna Be.* Dell, 1997.

130 Michael Jordan notes that he began dunking as a 5'8" freshman in the video *Come Fly with Me* (Fox/NBA), and his brother's athleticism and diminutive stature is often recounted, perhaps most eloquently in chapter 2 of David Halberstam's *Playing for Keeps: Michael Jordan and the World He Made.* Three Rivers Press, 2000.

131 Gene mixing may be contributing to widespread increase in height:

Malina, Robert M. (1979). "Secular Changes in Size and Maturity: Causes and Effects." *Monographs of the Society for Research in Child Development*, 44(3/4): 59–102.

131 Scientific papers addressing the threshold claims of journalists, including Malcolm Gladwell and David Brooks:

Arneson, Justin J., Paul R. Sackett, and Adam S. Beatty (2011). "Ability-Performance Relationships in Education and Employment Settings: Critical Tests of the More-Is-Better and the Good-Enough Hypotheses." *Psychological Science*, 22(10):1336-42.

Hambrick, David Z., and Elizabeth J. Meinz (2011). "Limits on the Predictive Power of Domain-Specific Experience and Knowledge in Skilled Performance." *Current Directions in Psychological Science*, 20(5):275–79.

(The paper notes: children scoring in the 99.9th percentile on the SAT's math section by age thirteen are eighteen times more likely to get a math or science Ph.D. than children who "only" scored in the 99.1th percentile.)

131 Data analysis of NBA body types in this chapter is original, carried out by the author and psychologist Drew H. Bailey. We used data from the NBA combine and from U.S. government sources that are noted in the text.

133 The 5'3" Muggsy Bogues could dunk:
Foreman, Tom Jr. "Bogues, Webb Make Case for the Little Guy." Associated Press, February 16, 1985.

135 A fascinating account of the "creation" of Yao Ming:
Larmer, Brook. *Operation Yao Ming: The Chinese Sports Empire, American Big Business, and the Making of an NBA Superstar.* Gotham, 2005.

136 Average height of a seventeenth-century Frenchman:
Blue, Laura. "Why Are People Taller Today Than Yesterday?" *Time,* July 8, 2008.

136 J. M. Tanner's *Fetus into Man* (Harvard University Press, 1990) served as a source on growth trends in the industrialized world. It is where he recounts: the tale of the identical twin brothers raised in starkly different environments (p. 121); the growth patterns of twins (p. 123); that man did not evolve with the supermarket (p. 130); the leg length disparities between socioeconomic classes (p. 131); work indicating that blind children have distinct growth patterns (p. 146); and rapid leg growth during Japan's "economic miracle" (p. 159).

136 The study that accounted for 45 percent of the variance in height with DNA variations also discusses the general finding that height is about 80 percent heritable in a given population:
Yang, Jian, et al. (2010). "Common SNPs Explain a Large Proportion of the Heritability for Human Height." *Nature Genetics,* 42(7):565–69.

137 On the inability to find height genes:
Maher, Brendan (2008). "The Case of the Missing Heritability." *Nature,* 456: 18–21.

137 Female gymnasts delay menarche, but attain normal adult stature:
Norton, Kevin, and Tim Olds. *Anthropometrica.* UNSW Press, 2004, p. 313.

138 Leg length—and particularly leg growth in Japan—is also discussed in:
Eveleth, Phyllis B., and James M. Tanner. *Worldwide Variation in Human Growth* (2nd ed.). Cambridge University Press, 1991.

138 Charts of leg length by ethnicity:
Eveleth, Phyllis B., and James M. Tanner. "Chapter 9: Genetic Influence on Growth: Family and Race Comparisons." *Worldwide Variation in Human Growth* (2nd ed.). Cambridge University Press, 1990.

138 The 1968 Mexico City Olympic study (the quote regarding "persistent" ethnic differences appears on p. 73):
de Garay, Alfonso L., Louise Levine, and J. E. Lindsay Carter, eds. *Genetic and Anthropological Studies of Olympic Athletes.* Academic Press, 1974.

140 The original "Allen's rule" paper:

Allen, Joel Asaph (1877). "The Influence of Physical Conditions in the Genesis of Species." *Radical Review*, 1:108–140.

140 A massive body of research has extended Allen's and Bergmann's rules to humans. For one recent discussion and a listing of confirmatory studies:
Cowgill, Libby W., et al. (2012). "Development Variation in Ecogeographic Body Proportions." *American Journal of Physical Anthropology*, 148:557–70.

140 The 1998 analysis of body proportions in native populations around the world:
Katzmarzyk, Peter T., and William R. Leonard (1998). "Climatic Influences on Human Body Size and Proportions: Ecological Adaptations and Secular Trends." *American Journal of Physical Anthropology*, 106:483–503.

141 The 2010 "belly button" study:
Bejan, A., Edward C. Jones, and Jordan D. Charles (2010). "The Evolution of Speed in Athletics: Why the Fastest Runners Are Black and Swimmers White." *International Journal of Design & Nature*, 5(3):199–211.
Duke press release: "For Speediest Athletes, It's All in the Center of Gravity." July 12, 2010.

9

We Are All Black (Sort Of)

Race and Genetic Diversity

142 Background on the "Out of Africa" hypothesis and previously competing hypotheses:
Klein, Richard G. "Chapter 7: Anatomically Modern Humans." *The Human Career: Human Biological and Cultural Origins* (2nd ed.). University of Chicago Press, 1999.

143 One example of the human "family tree" diagram:
Tishkoff, Sarah A., and Kenneth K. Kidd (2004). "Implications of Biogeography of Human Populations for 'Race' and Medicine." *Nature Genetics*, 36(11): S21–27.

144 The intrepid band of our ancestors that left Africa was a small group:
Macaulay, V., et al. (2005). "Single, Rapid Coastal Settlement of Asia Revealed by Analysis of Complete Mitochondrial Genomes." *Science*, 308:1034–36.
Wade, Nicholas. "To People the World, Start with 500." *New York Times*, November 11, 1997, p. F1.

144 Molecular dating and fossil methods for the timing of the human-chimp split and the Out-of-Africa migration:
Gibbons, Ann (2012). "Turning Back the Clock: Slowing the Pace of Prehistory." *Science*, 338:189–91.

144 A succinct look at how genetic diversity decreases with distance from Africa:

Prugnolle, Franck, Andrea Manica, and François Balloux (2005). "Geography Predicts Neutral Genetic Diversity of Human Populations." *Current Biology*, 15(5):R159–60. See fig. 2.

146 Kenneth Kidd's coauthored CYP2E1 paper is an example of his rainbow diagrams that describe genetic diversity:
Lee, M. Y., et al. (2008). "Global Patterns of Variation in Allele and Haplotype Frequencies and Linkage Disequilibrium Across the CYP2E1 Gene." *The Pharmacogenomics Journal*, 8(5):349–56.

147 An excellent and accessible talk by Sarah Tishkoff on the genetic changes that allowed adult lactose digestion:
http://www.youtube.com/watch?v=sgNEb0itPOs.

148 Adult lactose intolerance is common in Rwanda:
Cox, Joseph A., and Francis G. Elliott (1974). "Primary Adult Lactose Intolerance in the Kivu Lake Area: Rwanda and the Bushi." *American Journal of Digestive Diseases*, 19(8):714–724.

148 A common gene variant confers immunity from a sports doping test:
Schulze, Jenny Jakobsson, et al. (2008). "Doping Test Results Dependent on Genotype of Uridine Diphospho-Glucuronosyl Transferase 2B17, the Major Enzyme for Testosterone Glucuronidation." *Journal of Clinical Endocrinology & Metabolism*, 93(7):2500–2506.

148 An interesting albeit technical paper on the 99.5 percent DNA similarity of humans:
Levy, Samuel, et al. (2007). "The Diploid Genome Sequence of an Individual Human." *PLoS Biology*, 5(10):e254.

149 The 2007 scientific breakthrough of the year, "human genetic variation":
Pennisi, Elizabeth (2007). "Breakthrough of the Year: Human Genetic Variation." *Science*, 318:1842–43.

149 Local ancestry of Iceland residents identifiable with DNA:
Helgason, A., et al. (2005). "An Icelandic Example of the Impact of Population Structure on Association Studies." *Nature Genetics*, 37(1):90–95.

149 DNA pinpoints European ancestry to within a few hundred miles:
Novembre, John, et al. (2008). "Genes Mirror Geography Within Europe." *Nature*, 456(7218):98–101.

149 A computer blindly grouped DNA into major geographic regions:
Rosenberg, Noah A., et al. (2002). "Genetic Structure of Human Populations." *Science*, 298(5602):2381–85.

149 The Stanford-led study of self-identified race and genetics:
Tang, Hua, et al. (2005). "Genetic Structure, Self-Identified Race/Ethnicity, and Confounding in Case-Control Association Studies." *American Journal of Human Genetics*, 76(2):268–75.

150 The Stanford press release ("Racial Groupings Match Genetic Profiles, Stanford Study Finds") for the study can be found here:
http://med.stanford.edu/news_releases/2005/january/racial-data.htm.

150   On skin color, UV radiation, and latitude:
Jablonski, Nina G., and George Chaplin (2000). "The Evolution of Human
Skin Coloration." *Journal of Human Evolution*, 39:57–106.

150   The main genetic and geographic clusters of people do "correlate with the
common concept of 'races'":
Tishkoff, Sarah A., and Kenneth K. Kidd (2004). "Implications of Biogeography of Human Populations for 'Race' and Medicine." *Nature Genetics*, 36(11):
S21–27.

150   The genetic backgrounds of African Americans:
Tishkoff, Sarah A., et al. (2009). "The Genetic Structure and History of Africans and African Americans." *Science*, 324(5930):1035–44.

150   Tishkoff's "little genetic differentiation" quote can be found in a University of
Pennsylvania press release:
http://www.upenn.edu/pennnews/current/node/3643.

151   The National Human Genome Research Institute on race, genetics, and geno-
typic and phenotypic diversity:
Race, Ethnicity and Genetics Working Group of the National Human Genome Research Institute (2005). "The Use of Racial, Ethnic, and Ancestral
Categories in *Human Genetics* Research." *American Journal of Human Genetics*,
77:519–32.

153   The original ACTN3 paper:
North, Kathryn N., et al. (1999). "A Common Nonsense Mutation Results
in α-Actinin-3 Deficiency in the General Population." *Nature Genetics*, 21:
353–54.

155   The first paper that documented a difference in ACTN3 variant frequency in
sprinters and the general population:
Yang, Nan, et al. (2003). "ACTN3 Genotype Is Associated with Human Elite
Athletic Performance." *American Journal of Human Genetics*, 73:627–31.

155   ACTN3 and athletic performance studies in populations around the world:
Eynon, Nir, et al. (2012). "The ACTN3 R577X Polymorphism Across Three
Groups of Elite Male European Athletes." *PLoS ONE*, 7(8):e43132.
Niemi, A. K., and K. Majamaa (2005). "Mitochondrial DNA and ACTN3 Geno-
types in Finnish Elite Endurance and Sprint Athletes." *European Journal of Hu-
man Genetics*, 13:965–69.
Papadimitriou, I. D., et al. (2008). "The ACTN3 Gene in Elite Greek Track and
Field Athletes." *International Journal of Sports Medicine*, 29:352–55.
Scott, Robert A., et al. (2010). "ACTN3 and ACE Genotypes in Elite Jamaican
and US Sprinters." *Medicine & Science in Sports & Exercise*, 42(1):107–12.
Yang, Nan, et al. (2007). "The ACTN3 R577X Polymorphism in East and West
African Athletes." *Medicine & Science in Sports & Exercise*, 39(11):1985–88.
ACTN3 data from Japanese sprinters was generously shared by Noriyuki Fuku
and Eri Mikami during a visit to the Department of Genomics for Longevity
and Health at the Tokyo Metropolitan Institute of Gerontology.

155   The spread of the ACTN3 X variant in humans may have been an evolutionary adaptation:

North, Kathryn (2008). "Why Is α-Actinin-3 Deficiency So Common in the General Population? The Evolution of Athletic Performance." *Twin Research and Human Genetics*, 11(4):384–94.

155   The best review of ACTN3 research and the impacts on muscle properties of α-actinin-3 deficiency:

Berman, Yemima, and Kathryn N. North (2010). "A Gene for Speed: The Emerging Role of α-Actinin-3 in Muscle Metabolism." *Physiology*, 25:250–59.

155   The idea that the ACTN3 X variant may have spread as an adaptation to agriculture is posited on p. 117 of:

Cochran, Gregory, and Henry Harpending. *The 10,000 Year Explosion: How Civilization Accelerated Human Evolution*. Basic Books, 2010.

10

The Warrior-Slave Theory of Jamaican Sprinting

159   An overview of theories of Jamaican sprint success (p. 2 has ACTN3 data for Jamaicans and other populations):

Irving, Rachael, and Vilma Charlton eds. *Jamaican Gold: Jamaican Sprinters*. University of the West Indies Press, 2010.

161   Lists of sprinters of Jamaican descent who compete for other countries and of Jamaican sprinters from Trelawny can be found in the annex of:

Robinson, Patrick. *Jamaican Athletics: A Model for 2012 and the World*. Black Amber, 2009.

(These are merely partial lists. The Trelawny list, for example, does not include Olympic 100-meter finalist Michael Green or 4×100-meter world champion Merlene Frazer, both of whom were born in Trelawny.)

163   A thorough history of Jamaica's Maroons (the "born Heroes" and "elevation of the soul" quotes appears on p. 45):

Campbell, Mavis C. *The Maroons of Jamaica 1655–1796*. Africa World Press, 1990.

163   A history of Jamaica, written with particular attention to the African-Jamaican perspective:

Sherlock, Philip, and Hazel Bennett. *The Story of the Jamaican People*. Ian Randle Publishers, 1998.

(The "dangerous inmates" quote and William Beckford's description of a cane fire appear on p. 134 and the "dare not" quote on p. 139. Descriptions of Maroon battles for independence and of Cudjoe and Nanny can be found in chapter 13: "The African-Jamaican Liberation Wars, 1650–1800.")

164 A fascinating contemporary history of the Maroons is in the unabridged re-
prints of early-nineteenth-century letters:
Dallas, Robert C. *The History of the Maroons: From Their Origin to the Establishment
of Their Chief Tribe at Sierra Leone* (vols. I and II). Adamant Media Corporation,
2005. (Originally published in 1803 by T. N. Longman and O. Rees.)

166 A description of the slave/warrior/sprinter story, with Michael Johnson's quote
from the Channel 4 documentary:
Beck, Sally. "Survival of the Fastest: Why Descendants of Slaves Will Take the
Medals in the London 2012 Sprint Finals." *Daily Mail,* June 30, 2012.

167 Y chromosomes of Jamaican men:
Benn Torres, Jada (2012). "Y Chromosome Lineages in Men of West African
Descent." *PLoS ONE,* 7(1):e29687.

167 Genetic studies of the demographics of Jamaica, with both Errol Morrison and
Yannis Pitsiladis as coauthors:
Deason, Michael L., et al. (2012). "Interdisciplinary Approach to the Demogra-
phy of Jamaica." *BMC Evolutionary Biology,* 12:24.
Deason, M., et al. (2012). "Importance of Mitochondrial Haplotypes and Maternal Lineage in Sprint Performance Among Individuals of West African Ancestry." *Scandinavian Journal of Medicine & Science in Sports,* 22:217–23.

167 DNA shows that Taino Native Americans did not die out in Jamaica. The study
also gives data on the degree of genetic "African-ness" of various Caribbean
populations:
Benn Torres, J., et al. (2007). "Admixture and Population Stratification in African Caribbean Populations." *Annals of Human Genetics,* 72:90–98.

169 A visit to Champs should be on the bucket list of any track-and-field fan. The
next best treat:
Lawrence, Hubert. *Champs 100: A Century of Jamaican High School Athletics, 1910–
2010.* Great House, 2010.

174 Pitsiladis's advice to prospective white sprinters appears here:
"No Proof Sporting Success Is Genetic According to Academic." Scotsman
.com, March 23, 2011.

11

Malaria and Muscle Fibers

175 Background on latitude and pelvic breadth:
Nuger, Rachel Leigh. *The Influence of Climate on the Obstetrical Dimensions of the
Human Bony Pelvis.* UMI Dissertation Publishing, 2011.

175 The Cooper and Morrison paper introducing their hypothesis:
Morrison, E. Y. St. A., and P. D. Cooper (2006). "Some Bio-Medical Mechanisms in Athletic Prowess." *West Indian Medical Journal,* 55(3):205–209.

176   Patrick Cooper's widow Juin—and several obits—provided details of his life. Cooper's book on black athletes:
Cooper, Patrick Desmond. *Black Superman: A Cultural and Biological History of the People That Became the World's Greatest Athletes.* First Sahara, 2003.

177   The famous study of 1968 Mexico City Olympians, again:
de Garay, Alfonso L., Louise Levine, and J. E. Lindsay Carter, eds. *Genetic and Anthropological Studies of Olympic Athletes.* Academic Press, 1974.

177   Underrepresentation of sickle-cell carriers at race distances of eight hundred meters and above:
Eichner, Randy E. (2006). "Sickle Cell Trait and the Athlete." *Gatorade Sports Science Institute: Sports Science Exchange,* 19(4):103.

177   Analysis of the risk of death to college football players with sickle-cell trait:
Harmon, Kimberly G., et al. (2012). "Sickle Cell Trait Associated with a RR of Death of 37 Times in National Collegiate Athletic Association Football Athletes: A Database with 2 Million Athlete-Years as Denominator." *British Journal of Sports Medicine,* 46:325–30.

178   The first article Cooper cited showing low hemoglobin levels in African Americans:
Garn, Stanley M., Nathan J. Smith, and Diance C. Clark (1975). "Lifelong Differences in Hemoglobin Levels Between Blacks and Whites." *Journal of the National Medical Association,* 67(2):91–96.

178   Data tables from the CDC's National Center for Health Statistics are publicly available, and are easily located with a call to the Center. Heaps of hemoglobin data are also available in published reports:
Hollowell J. G., et al. (2005). "Hematological and Iron-Related Analytes— Reference Data for Persons Aged 1 Year and Over: United States, 1988–94." National Center for Health Statistics. *Vital Health Statistics,* 11(247).
Robins, Edwin B., and Steve Blum (2007). "Hematologic Reference Values for African American Children and Adolescents." *American Journal of Hematology,* 82:611–14.

178   Study of 715,000 blood donors:
Mast, Alan E., et al. (2010). "Demographic Correlates of Low Hemoglobin Deferral Among Prospective Whole Blood Donors." *Transfusion,* 50(8): 1794–1802.

179   The quote in which doctors refer to "some compensatory mechanism" appears here:
Kraemer, Michael J., et al. (1977). "Race-Related Differences in Peripheral Blood and in Bone Marrow Cell Populations of American Black and American White Infants." *Journal of the National Medical Association,* 69(5):327–31.

179   The fiber type study coauthored by Bouchard:
Ama, P. F., et al. (1986). "Skeletal Muscle Characteristics in Sedentary Black and Caucasian Males." *Journal of Applied Physiology,* 61(5):1758–61.

180 Sickle-cell trait causes reduced capacity to produce energy through pathways that rely primarily on oxygen:
Bitanga, E., and J. D. Rouillon (1998). "Influence of the Sickle Cell Trait Heterozygote on Energy Abilities." *Pathologie Biologie*, 46(1):46–52.
Le Gallais, D., et al. (1994). "Sickle Cell Trait as a Limiting Factor for High-Level Performance in a Semi-Marathon." *International Journal of Sports Medicine*, 15(7):399–402.

180 For quick background on the malaria protection conferred by sickle-cell trait:
Pierce, E. C. "How Sickle Cell Trait Protects Against Malaria." *Medical Journal of Therapeutics Africa*, 1(1):61–62.

180 Anthony C. Allison first documented the connection between sickle-cell trait and malaria resistance:
Allison, A. C. (1954). "Protection Afforded by Sickle-Cell Trait Against Subtertian Malarial Infection." *British Medical Journal*, 1(4857):290–94.
Allison, Anthony C. (2002). "The Discovery of Resistance to Malaria of Sickle-Cell Heterozygotes." *Biochemistry and Molecular Biology Education*, 30(5):279–87.

181 The gradual disappearance of the sickle-cell gene in African Americans is discussed on p. 99 of:
Nesse, Randolph M., and George C. Williams. *Why We Get Sick: The New Science of Darwinian Medicine*. Vintage, 1996.

181 Risk of malaria with iron supplementation has long been documented by Stephen J. Oppenheimer and others:
English, M., and R. W. Snow (2006). "Iron and Folic Acid Supplementation and Malaria Risk." *Lancet*, 367(9505):90–91.
Oppenheimer, S. J., et al. (1986). "Iron Supplementation Increases Prevalence and Effects of Malaria: Report on Clinical Studies in Papua New Guinea." *Transactions of the Royal Society of Tropical Medicine and Hygiene*, 80(4)603–12.
Oppenheimer, Stephen (2007). "Comments on Background Papers Related to Iron, Folic Acid, Malaria and Other Infections." *Food and Nutrition Bulletin*, 28(4):S550–59.

182 In 2006, the WHO revised iron supplementation recommendations for malaria zones: http://www.who.int/maternal_child_adolescent/documents/iron_statement/en/.

182 The global pattern of the sickle-cell gene and its relation to malaria (with color-coded maps available online):
Piel, Frédéric B., et al. (2010). "Global Distribution of the Sickle Cell Gene and Geographical Confirmation of the Malaria Hypothesis." *Nature Communications*, 1:104.

182 Danish scientists proposed that fast-twitch fibers might explain physical traits documented in African Americans:

Nielsen, J., and D. L. Christensen (2011). "Glucose Intolerance in the West African Diaspora: A Skeletal Muscle Fibre Type Distribution Hypothesis." *Acta Physiologica*, 202(4):605–16.

183   Daniel Le Gallais's coauthored studies on athletic performance and sickle-cell trait:

Bilé A., et al. (1998). "Sickle Cell Trait in Ivory Coast Athletic Throw and Jump Champions, 1956–1995." *International Journal of Sports Medicine*, 19(3):215–19.

Hue, O., et al. (2002). "Alactic Anaerobic Performance in Subjects with Sickle Cell Trait and Hemoglobin AA." *International Journal of Sports Medicine*, 23(3): 174–77.

Le Gallais, D., et al. (1994). "Sickle Cell Trait as a Limiting Factor for High-Level Performance in a Semi-Marathon." *International Journal of Sports Medicine*, 15(7):399–402.

Marlin, L., et al. (2005). "Sickle Cell Trait in French West Indian Elite Sprint Athletes." *International Journal of Sports Medicine*, 26(8):622–25.

184   The two studies showing a muscle fiber type proportion shift in low hemoglobin mice:

Esteva, Santiago, et al. (2008). "Morphofunctional Responses to Anaemia in Rat Skeletal Muscle." *Journal of Anatomy*, 212:836–44.

Ohira, Yoshinobu, and Sandra L. Gill (1983). "Effects of Dietary Iron Deficiency on Muscle Fiber Characteristics and Whole-Body Distribution of Hemoglobin in Mice." *Journal of Nutrition*, 113:1811–18.

185   In populations at altitude in East Africa the sickle-cell mutation is rare or nonexistent:

Ayodo, George, et al. (2007). "Combining Evidence of Natural Selection with Association Analysis Increases Power to Detect Malaria-Resistance Variants." *American Journal of Human Genetics*, 81:234–42.

Foy, Henry, et al. (1954). "The Variability of Sickle-Cell Rates in the Tribes of Kenya and the Southern Sudan." *British Medical Journal*, 1(4857):294.

Williams, Dianne. Race, *Ethnicity and Crime: Alternate Perspectives*. Algora Publishing, 2012, p. 20.

12

Can Every Kalenjin Run?

186   A breakdown on who the elite runners in Kenya are and what tribes they come from:

Onywera, Vincent O., et al. (2006). "Demographic Characteristics of Elite Kenyan Endurance Runners." *Journal of Sports Sciences*, 24(4):415–22.

190   Cattle raiding was not regarded as theft so long as not from the same tribe:

Bale, John, and Joe Sang. *Kenyan Running: Movement Culture, Geography and Global Change*. Frank Cass, 1996, p. 53.

190　The best compilation of scholarly writing examining the success of East African runners:

Pitsiladis, Yannis, et al., eds. *East African Running: Towards a Cross-Disciplinary Perspective*. Routledge, 2007.

190　Ethiopian population data comes from the "Summary and Statistical Report of the 2007 Population and Housing Census," issued by Ethiopia's Public Census Commission.

190　John Manners's writing about the "cattle complex" and a number of his accounts of Kalenjin phenoms and his written quotes on Rotich:

Manners, John (1997). "Kenya's Running Tribe." *The Sports Historian*, 17(2):14–27.

Manners, John. "Chapter 3: Raiders from the Rift Valley: Cattle Raiding and Distance Running in East Africa." In: Yannis Pitsiladis, et al., eds. *East African Running: Towards a Cross-Disciplinary Perspective*. Routledge, 2007.

193　From the IAAF list of top marathon times of 2011; John Manners assisted in identifying the Kalenjin athletes.

194　Scott Bickard's comparison of Peter Kosgei to an NBA player appeared in the *Utica Observer-Dispatch* on April 21, 2011.

195　A succinct summary of the Copenhagen research team's work—including Saltin's "seems to confirm" quote:

Saltin, Bengt (2003). "The Kenya Project—Final Report." *New Studies in Athletics*, 18(2):15–24.

195–98　A more technical description is here:

Larsen, Henrik B. (2003). "Kenyan Dominance in Distance Running." *Comparative Biochemistry and Physiology Part A: Molecular & Integrative Physiology*, 136(1):161–70.

197　Another study finding that African distance runners have better economy at a given pace than white runners:

Weston, A. R., Z. Mbambo, and K. H. Myburgh (2000). "Running Economy of African and Caucasian Distance Runners." *Medicine & Science in Sports & Exercise*, 32(6):1130–34.

197　Distal weight and running energetics (what happens when weight is added to the ankle):

Jones, B. H. et al. (1986). "The Energy Cost of Women Walking and Running in Shoes and Boots." *Ergonomics*, 29:439–43.

Myers, M. J., and K. Steudel (1985). "Effect of Limb Mass and Its Distribution on the Energetics Cost of Running." *Journal of Experimental Biology*, 116:363–73.

197　Harvard's Dan Lieberman also confirmed the increased energetic cost of distal weight, and the finding by Adidas engineers was conveyed to me by Andrew Barr, global product line manager for Adidas running products.

197 Longer legs and thinner lower legs contribute separately to running economy:
Steudel-Numbers, Karen L., Timothy D. Weaver, and Cara M. Wall-Scheffler (2007). "The Evolution of Human Running: Effects of Changes in Lower-Limb Length on Locomotor Economy." *Journal of Human Evolution*, 53(2):191–96.

197 Kenyan runners and their long Achilles tendons:
Sano, K., et al. (2012). "Muscle-Tendon Interaction and EMG Profiles of World Class Endurance Runners During Hopping." *European Journal of Applied Physiology*, December 11 (ePub ahead of print).

198 Larsen's contention that the main point of Kenyan running dominance has been solved appears here:
Holden, Constance (2004). "Peering Under the Hood of Africa's Runners." *Science*, 305(5684):637–39.

198 Zersenay Tadese's running economy:
Lucia, Alejandro, et al. (2007). "The Key to Top-Level Endurance Running Performance: A Unique Example." *British Journal of Sports Medicine*, 42:172–174.

201 Vincent Sarich's calculation starts on p. 174 of:
Sarich, Vincent, and Frank Miele. *Race: The Reality of Human Differences*. Westview Press, 2004.

202 The Runner's World calculation appears in:
Burfoot, Amby (1992). "White Men Can't Run." *Runner's World*, 27(8):89–95.

13

The World's Greatest Accidental (Altitudinous) Talent Sieve

207 Most Kenyan runners are Kalenjin and traveled to school on foot:
Onywera, Vincent O., et al. (2006). "Demographic Characteristics of Elite Kenyan Endurance Runners." *Journal of Sports Science*, 24(4):415–22.

208 Most Ethiopian runners are Oromo and traveled to school on foot:
Scott, Robert A., et al. (2003). "Demographic Characteristics of Elite Ethiopian Endurance Runners." *Medicine & Science in Sports & Exercise*, 35(10):1727–32.

208 Mitochondrial DNA of Oromo Ethiopians and Kalenjin Kenyans is not particularly closely related:
Scott, Robert A., et al. (2008). "Mitochondrial Haplogroups Associated with Elite Kenyan Athlete Status." *Medicine & Science in Sports & Exercise*, 41(1):123–28.
Scott, Robert A., et al. (2005). "Mitochondrial DNA Lineages of Elite Ethiopian Athletes." *Comparative Biochemistry and Physiology Part B: Biochemistry and Molecular Biology*, 140(3):497–503.

211   Nineteenth-century scientists were unaware of the variety of altitude adaptation Beall would find:
      Beall, Cynthia M. (2006). "Andean, Tibetan, and Ethiopian Patterns of Adaptation to High-Altitude Hypoxia." *Integrative and Comparative Biology*, 46(1):18–24.

212   Beall raised the possibility that Ethiopians living at high altitude have enhanced transfer of oxygen from lungs to blood. (Snell's theorizing on that topic was directly to the author in an interview.):
      Beall, Cynthia M., et al. (2002). "An Ethiopian Pattern of Human Adaptation to High-Altitude Hypoxia." *Proceedings of the National Academy of Sciences*, 99(26):17215–18.

213   Kenenisa Bekele's altitude workout data was generously shared by Barry Fudge, senior physiologist at the English Institute of Sport.

214   Scientists from Norway and Texas exposed athletes to altitude and documented EPO changes:
      Jedlickova, K., et al. (2003). "Search for Genetic Determinants of Individual Variability of the Erythropoietin Response to High Altitude." *Blood Cells, Molecules & Diseases*, 31(2):175–82.

214   The response of red blood cell levels and 5K times to altitude is highly individual:
      Chapman, Robert F. (1998). "Individual Variation in Response to Altitude Training." *Journal of Applied Physiology*, 85(4):1448–56.

214   Information on the altitude "sweet spot" comes from numerous interviews with altitude experts, including Randall L. Wilber, senior sport physiologist at the U.S. Olympic Training Center in Colorado Springs, Colorado. A good background resource—including a list of the altitudes of famous training cities:
      Wilber, Randall L. *Altitude Training and Altitude Performance*. Human Kinetics, 2004.

215   Children who grow up at altitude have larger lung surface area, but adults who move there do not:
      Moore, Lorna G., Susan Niermeyer, and Stacy Zamudio (1998). "Human Adaptation to High Altitude: Regional and Life-Cycle Perspectives." *Yearbook of Physical Anthropology*, 41:25–64.

215   High-altitude Ethiopians have larger forced expiratory volume of airflow than Ethiopian lowlanders. (Also contains a table with some measures of stature and sitting height in Ethiopians):
      Harrison, G. A., et al. (1969). "The Effects of Altitudinal Variation in Ethiopian Populations." *Philosophical Transactions of the Royal Society of London. Series B, Biological Sciences*, 805(256):147–82.

220   Claudio Berardelli's coauthored paper on the running economy of European and Kenyan runners:
      Tam, E., et al. (2012). "Energetics of Running Top-Level Marathon Runners from Kenya." *European Journal of Applied Physiology*, 112(11):3797–806.

221   Andrew M. Jones's years of physiological testing on Paula Radcliffe:

Jones, Andrew M. (2006) "The Physiology of the World Record Holder for the Women's Marathon." *International Journal of* Sports *Science* & Coaching, 1(2):101–16.

222 Sir Roger Bannister's quote appeared in the June 20, 1955, issue of *Sports Illustrated*.

14

Sled Dogs, Ultrarunners, and Couch Potato Genes

223 A candid and engrossing account of Lance Mackey's life, in his own words: Mackey, Lance. *The Lance Mackey Story: How My Obsession with Dog Mushing Saved My Life*. Zorro Books, 2010.

231 Physiologist and veterinarian Michael Davis (Oklahoma State University) gave an accessible talk on his research on the exercise adaptation of sled dogs at Texas A&M's Huffines Discussion 2012. (I was also an invited speaker, and had the pleasure of discussing Dr. Davis's work with him.) His talk can be found here: http://huffinesinstitute.org/resources/videos/entryid/330/huffines-discussion-2012-oklahoma-states-dr-michael-davis.

232 The genetics of Alaskan huskies:
Huson, Heather J., et al. (2010). "A Genetic Dissection of Breed Composition and Performance Enhancement in the Alaskan Sled Dog." *BMC Genetics*, 11:71.

234 Garland's coauthored work on dopamine, Ritalin, and "running-junkie" mice:
Rhodes, J. S., S. C. Gammie, and T. Garland Jr. (2005). "Neurobiology of Mice Selected for High Voluntary Wheel-Running Activity." *Integrative and Comparative Biology*, 45(3):438–55.

236 The University of Wisconsin mice to which Pam Reed compared herself:
Rhodes, J. S., T. Garland Jr., and S. C. Gammie (2003). "Patterns of Brain Activity Associated with Variation in Voluntary Wheel Running Behavior." *Behavioral Neuroscience*, 117(6):1243–56.

237 Background on the scientific study of dopamine and addiction:
Holden, Constance (2001). " 'Behavioral' Addictions: Do They Exist?" *Science*, 294:980–82.
Peirce, R. C., and V. Kumaresan (2006). "The Mesolimbic Dopamine System: The Final Common Pathway for the Reinforcing Effect of Drugs of Abuse?" *Neuroscience & Biobehavioral Reviews*, 30(2):215–38.

238 Every human study conducted has found that voluntary physical activity is significantly heritable:
Lightfoot, J. Timothy (2011). "Current Understanding of the Genetic Basis for Physical Activity." *Journal of Nutrition*, 141(3):526–30.

238 In thirteen thousand Swedish twin pairs, identical twins were far more likely to be similarly active or inactive:

Carlsson, S., et al. (2006). "Genetic Effects on Physical Activity: Results from the Swedish Twin Registry." *Medicine & Science in Sports & Exercise*, 38(8):1396–1401.

238   When activity is directly measured with accelerometers, the difference between fraternal and identical twins holds:
Joosen, A. M., et al. (2005). "Genetic Analysis of Physical Activity in Twins." *American Journal of Clinical Nutrition*, 82(6):1253–59.

238   Stubbe, Janine H., et al. (2006). "Genetic Influences on Exercise Participation in 37,051 Twin Pairs from Seven Countries." *PLoS ONE*, 1:e22.

239   Review of research on the dopamine system—and early work on genes—and voluntary physical activity:
Knab, Amy M., and J. Timothy Lightfoot (2010). "Title: Does the Difference Between Physically Active and Couch Potato Lie in the Dopamine System?" *International Journal of Biological Science*, 6(2):133–50.

239   DRD4-7R and ADHD:
Li, D., et al. (2006). "Meta-analysis Shows Significant Association Between Dopamine System Genes and Attention Deficit Hyperactivity Disorder (ADHD)." *Human Molecular Genetics*, 15(14):2276–84.
Swanson, J. M., et al. (2007). "Etiologic Subtypes of Attention-Deficit/Hyperactivity Disorder: Brain Imaging, Molecular Genetic and Environmental Factors and the Dopamine Hypothesis." *Neuropsychology Review*, 17(1):39–59.

240   The DRD4 gene in migratory and settled cultures:
Chen, Chuansheng, et al. (1999). "Population Migration and the Variation in Dopamine D4 Receptor (DRD4) Allele Frequencies Around the Globe." *Evolution and Human Behavior*, 20:309–24.
Matthews, L. J., and P. M. Butler (2011). "Novelty-Seeking DRD4 Polymorphisms Are Associated with Human Migration Distance Out-of-Africa After Controlling for Neutral Population Gene Structure." *American Journal of Physical Anthropology*, 145(3):382–89.

240   The DRD4 gene and Ariaal tribesmen:
Eisenberg, Dan T. A., et al. (2008). "Dopamine Receptor Genetic Polymorphisms and Body Composition in Undernourished Pastoralists: An Exploration of Nutrition Indices Among Nomadic and Recently Settled Ariaal Men of Northern Kenya." *BMC Evolutionary Biology*, 8:173.

15

The Heartbreak Gene

Death, Injury, and Pain on the Field

245   The best background resources on sudden death in athletes:
Estes III, Mark N. A., Deeb N. Salem, and Paul J. Wang, eds. *Sudden Cardiac Death in the Athlete*. Futura, 1998.

Maron, Barry J., ed. *Diagnosis and Management of Hypertrophic Cardiomyopathy*. Futura, 2004.

245 In my *Sports Illustrated* article "Following the Trail of Broken Hearts" (December 10, 2007), I made the analogy of an HCM mutation to a typo in *Encyclopaedia Britannica*. There, I analogized a single DNA base change to one typo in sixty full sets of *Britannica*. In this book I used thirteen full sets of *Britannica*. In *SI*, I counted each word in the *Encyclopaedia Britannica* set as a possible single typo. In this book, I considered each individual letter as a possible typo—a scenario that I think is more accurately compared to DNA.

245 An excellent primer on HCM, written specifically for the layman and with pictures of the heart cells:
Maron, Barry J., and Lisa Salberg. *Hypertrophic Cardiomyopathy: For Patients, Their Families and Interested Physicians* (2nd ed.). Wiley-Blackwell, 2006.

247 The MYH7 gene was the first, but now many mutations that cause HCM have been identified:
Maron, Barry J., Martin S. Maron, and Chrisopher Semsarian (2012). "Genetics of Hypertrophic Cardiomyopathy After 20 Years." *Journal of the American College of Cardiology*, 60(8):705–15.

248 The weight of Kevin Richards's heart comes from his autopsy documents, obtained with written permission of his parents, Gwendolyn and Rupert Richards.

248 An increasing number of states allow nonphysicians to conduct preparticipation exams:
Glover, David W., Drew W. Glover, and Barry J. Maron (2007). "Evolution in the Process of Screening United States High School Student-Athletes for Cardiovascular Disease." *American Journal of Cardiology*, 100:1709–12.

251 Alan Milstein's quote originally appeared here:
Litke, Jim. "Curry's DNA Fight with Bulls 'Bigger Than Sports World.'" Associated Press, September 29, 2005.

254 ApoE4 carriers get Alzheimer's more often and younger:
Corder, E. H., et. al. (1993). "Gene Dose of Apolipoprotein E type 4 Allele and the Risk of Alzheimer's Disease in Late Onset Families." *Science*, 261(5123):921–23.

254 ApoE4 influences severity of brain trauma injury:
Jordan, Barry D. (2007). "Genetic Influences on Outcome Following Traumatic Brain Injury." *Neurochemical Research*, 32:905–15.

254 Boxers with ApoE4 have worse outcomes:
Jordan, Barry D. (1997). "Apolipoprotein E epsilon4 Associated with Chronic Traumatic Brain Injury." *Journal of the American Medical Association*, 278(2):136–40.

254 Age, getting hit in the head, and ApoE4 negatively influence brain function:
Kutner, K. C., et al. (2000). "Lower Cognitive Performance of Older Football Players Possessing Apolipoprotein E epsilon4." *Neurosurgery*, 47(3):651–57.

255   BU's Center for the Study of Traumatic Encephalopathy has background
      on CTE, and John Grimsley's brain:
      http://www.bumc.bu.edu/supportingbusm/research/brain/cte/.

255   Two percent of people have two copies of the ApoE4 gene variant:
      Izaks, Gerbrand J., et al. (2011). "The Association of ApoE Genotype with
      Cognitive Function in Persons Aged 35 Years or Older." *PLoS ONE*,
      6(11):e27415.

255   BU researchers have been compiling cases of CTE in athletes:
      McKee, Ann C., et al. (2009). "Chronic Traumatic Encephalopathy in Ath-
      letes: Progressive Tauopathy Following Repetitive Head Injury." *Journal of
      Neuropathology & Experimental Neurology*, 68(7):709–35.

257   Sam Gandy, director of Mt. Sinai Hospital's Center for Cognitive Health,
      equated ApoE4 risk to playing in the NFL:
      http://www.alzforum.org/new/detail.asp?id=3264.

257   When people learn what version of ApoE they have:
      Green, Robert C., et al. (2009). "Disclosure of ApoE Genotype for Risk of
      Alzheimer's Disease." *New England Journal of Medicine*, 361:245–54.

257   Technical background on research into genes that may affect injury sus-
      ceptibility:
      Collins, Malcolm, and Stuart M. Raleigh. "Genetic Risk Factors for Muscu-
      loskeletal Soft Tissue Injuries." In: Malcolm Collins, ed. *Genetics and Sports*.
      Karger, 2009, 54:136–49.

258   COL5A1 may also influence flexibility and running performance via
      Achilles tendon stiffness:
      Posthumus, Michael, Martin P. Schwellnus, and Malcolm Collins (2011).
      "The COL5A1 Gene: A Novel Marker of Endurance Running Perfor-
      mance." *Medicine & Science in Sports & Exercise*, 43(4):584–89.

258   A number of NFL players have pursued "injury gene" testing:
      Assael, Shaun. "Cheating Is So 1999." *ESPN The Magazine*, October 8, 2009,
      pp. 88–97.

260   An excellent resource—but very technical—for a broad look at the pain
      genetics landscape:
      Mogil, Jeffrey S. *The Genetics of Pain*. IASP Press, 2004.

260   The "redhead" mutation reduces pain sensitivity:
      Mogil, J., et al. (2005). "Melanocortin-1 Receptor Gene Variants Affect
      Pain and μ-Opioid Analgesia in Mice and Humans." *Journal of Medical Ge-
      netics*, 42(7):583–87.

261   The quote from British researchers regarding a Pakistani family's inability
      to feel pain appears here:
      Cox, James J., et al. (2006). "An SCN9A Channelopathy Causes Congenital
      Inability to Experience Pain." *Nature*, 444(7121):894–98.

261   Pain perception is altered by common variation in SCN9A:

Reimann, Frank, et al. (2010). "Pain Perception Is Altered by a Nucleotide Polymorphism in SCN9A." *Proceedings of the National Academy of Sciences*, 107(11):5148–53.

261 Background on the COMT gene:
Goldman, David. "Chapter 13: Warriors and Worriers." *Our Genes, Our Choices: How Genotype and Gene Interactions Affect Behavior*. Academic Press, 2012.
Stein, Dan J., et al. (2006). "Warriors Versus Worriers: The Role of COMT Gene Variants." *Pearls in Clinical Neuroscience*, 11(10):745–48.

263 Athletes are less sensitive to pain on game day:
Sternberg, W. F., et al. (1998). "Competition Alters the Perception of Noxious Stimuli in Male and Female Athletes." *Pain*, 76(1–2):231–38.

16

The Gold Medal Mutation

274 The first documentation of the inheritance pattern of high red blood cell levels in the Mäntyranta family:
Juvonen, Eeva, et al. (1991). "Autosomal Dominant Erythrocytosis Caused by Increased Sensitivity to Erythropoietin." *Blood*, 78(11):3066–69.

276 First documentation of the Mäntyranta family EPOR mutation:
de la Chapelle, Albert, et al. (1993). "Familial Erythrocytosis Genetically Linked to Erythropoietin Receptor Gene." *Lancet*, 341:82–84.

277 Detailed analysis of the Mäntyranta family EPOR mutation:
de la Chapelle, Albert, Ann-Liz Träskelin, and Eeva Juvonen (1993). "Truncated Erythropoietin Receptor Causes Dominantly Inherited Benign Human Erythrocytosis." *Proceedings of the National Academy of Sciences*, 90:4495–99.

EPILOGUE

The Perfect Athlete

286 Williams, Alun G., and Jonathan P. Folland (2008). "Similiarity of Polygenic Profiles Limits the Potential for Elite Human Physical Performance." *The Journal of Physiology*, 586(pt. 1):113–21.

288 Cunningham, Patrick. "The Genetics of Thoroughbred Horses." *Scientific American* (May 1991).

Index

Index

Index

Index

Index

Index